Planetary Rock Mechanics and Rock Engineering with Emphasis on Mars

Planetary Rock Mechanics and Rock Engineering with Emphasis on Mars presents a pioneering exploration of rock mechanics principles applied to celestial bodies beyond Earth. Emerging from the author's establishment of the Planetary Rock Mechanics Commission within ISRM in 2019, this comprehensive work analyses spectacular imagery from Mars, the Moon and Venus rovers, plus spacecraft data from comets and asteroids, to develop theoretical frameworks for understanding extraterrestrial rock behavior and engineering challenges.

The book offers unique insights into planetary rock formations, weathering processes, discontinuity patterns, and slope stability phenomena across the Solar System. Key features include detailed Martian geological analysis, comparative studies of terrestrial planetary bodies, and practical applications for underground cavity stability and impact crater formation. Readers benefit from comprehensive coverage of fundamental laws adapted for planetary conditions, stress state analysis of celestial bodies, and seismic activity patterns beyond Earth. Special attention is given to similarities between Earth's Salda Lake and Mars's Jezero Crater, providing valuable geological analogues.

This essential resource serves rock mechanics and engineering communities, academic researchers in planetary science and geological engineering, graduate students, and industry professionals in space mining and exploration missions. The book appeals to interdisciplinary researchers studying celestial bodies for scientific, commercial, or colonization purposes, offering foundational knowledge in planetary material mechanics and extraterrestrial engineering challenges.

ISRM Book Series

Series editors: Xia-Ting Feng, *Institute of Rock and Soil Mechanics, Chinese Academy of Sciences, Wuhan, China,* and **Reşat Ulusay**, *Department of Geological Engineering, Hacettepe University, Ankara, Turkey*

International Society for Rock Mechanics and Rock Engineering

ISRM

The ISRM Book Series comprises significant scientific publications in rock mechanics and related disciplines. The aim of the Series is to promote the scientific output of ISRM. The Series is published on a regular basis, according to the emergence of noteworthy scientific developments. Each volume in the Series is prepared independently and focuses on a topical theme. Contributions are rigorously peer-reviewed by experts in the field, ensuring strong scientific standards, and are expected to be leading in rock mechanics and related disciplines. The Series may include edited volumes, textbooks and reference works.

Time-Dependency in Rock Mechanics and Rock Engineering
Ömer Aydan

Rock Dynamics
Ömer Aydan

Discontinuous Deformation Analysis in Rock Mechanics Practice
Yossef H. Hatzor, Guowei Ma, and Gen-hua Shi

Back Analysis in Rock Engineering
Shunsuke Sakurai

Rock Reinforcement and Rock Support
Ömer Aydan

Continuum and Computational Mechanics for Geomechanical Engineers
Ömer Aydan

Storage of LPG in Large Rock Caverns
Kenji Aoki

Geomechanical Aspects of Abandoned Room and Pillar Mines and Remediation Measures
Ömer Aydan

Planetary Rock Mechanics and Rock Engineering with Emphasis on Mars
Ömer Aydan

For more information about this series, please visit: www.routledge.com/ISRM-Book-Series/book-series/ISRM

Planetary Rock Mechanics and Rock Engineering with Emphasis on Mars

Ömer Aydan

Emeritus Professor, University of the Ryukyus, Okinawa, Japan

CRC Press
Taylor & Francis Group
Boca Raton London New York

CRC Press is an imprint of the
Taylor & Francis Group, an **informa** business

Designed cover image: Top-left: Gediz Vallis in Mars (NASA), Top-right: crater in sandstone by a ballistic missile (picture by Aydan), Middle-left: Mankind standing to Lunar rock boulder (Apollo 17), Middle-right: Partially collapsed lava tune in Karacadağ, Türkiye, Bottom-left drilling in Yonabaru mudstone of Okinawa (picture by Aydan), Bottom-center: Planar sliding in Mars by Curiosity rover, Bottom-right: Comet Hale Bopp (Courtesy of Prof. Dr. H. Tano, Koriyama, Japan)

First edition published 2027
by CRC Press
4 Park Square, Milton Park, Abingdon, Oxon, OX14 4RN

and by CRC Press
2385 NW Executive Center Drive, Suite 320, Boca Raton FL 33431

CRC Press is an imprint of informa UK Limited

British Library Cataloguing-in-Publication Data
A catalogue record for this book is available from the British Library

ISBN: 978-1-041-08122-7 (hbk)
ISBN: 978-1-041-08125-8 (pbk)
ISBN: 978-1-003-64385-2 (ebk)

DOI: 10.1201/9781003643852

Typeset in Times
by Apex CoVantage, LLC

Contents

Preface

There is recently an increasing interest to explore the ways to find out the characteristics of other planets and the possibility of exploiting their mineral resources. The USA and Russia (former SSCB) sent several spacecraft to the Moon, Venus and Mars, and some of them landed on them. NASA released images taken by Mars exploration rovers (MERs) that showed the striking similarities between the rocks on Earth and those of Mars. One can easily notice layered rock masses, jointing, weathering effects and slope stability issues such as plane and wedge sliding, block and flexural toppling-type failures.

The dynamics issues on Mars, the Moon, Venus and other planetary objects should be similar to those on Earth. Although humankind has no direct access to Mars yet, the Mars exploration rovers (e.g., Opportunity, Spirit, Curiosity and Perseverance), the Mars Reconnaissance Orbiter and InSight Lander have provided very valuable data so that some dynamics issues can be recognized.

Low-gravity planetary bodies such as comets and asteroids have been of great interest to humankind since early times. The recent attempt by the European Space Agency (ESA) on the Comet 67P/Churyumov–Gerasimenko revealed a lot of information on the various strengths of cometary material as well some familiar phenomena of slope failures, sinkholes, debris flow and boulder toppling in a low-gravitational environment. Attempts to bring materials from several asteroids were recently successful, and they may provide important information on the characteristics and how to mine the asteroids.

Some of most impressive images from the Apollo Program of NASA are the astronaut next to a lunar rock mass and hitting lunar rock with a geologist's hammer, which motivated the author to be interested in the rock mechanics of planetary bodies. The author also felt the necessity to teach Planetary Rock Mechanics in his Rock Mechanics Classes in Tokai University since 2010 and at the University of the Ryukyus since 2014. This was also another step to try to initiate learning and teaching activities in academia.

The author was invited by Prof. S. Saydam to give a talk on "Some Thoughts about Rock Mechanics Aspects of Mars" at the 3rd Off Earth Mining Forum—2017OEMF held in the University of New South Wales (UNSW), Sydney, Australia, in 2017, which made the author really involved with Planetary Rock Mechanics. This encounter eventually led to establishing the Planetary Rock Mechanics Commission in ISRM in 2019. This commission has now entered the second period in 2024, and it is expected to promote the activities of the commission as well as the ISRM community.

This book covers the thoughts of the author on the aspects of planetary rock mechanics and rock engineering deducted from the images sent to the earth from Mars, the Moon and Venus by exploration rovers and from comets and asteroids via spacecraft. Particularly, based on the images from Mars, it is spectacular to imagine and infer many aspects of planetary rock mechanics and resemblance of rocks and their weathering, discontinuity formations, slope stability problems and natural caves, leading to humankind's contemplating the possible mechanics and material behavior of rocks and their related scientific and engineering aspects.

The book has 14 chapters. Chapter 1 briefly explains the motivation as well as a summary of planetary bodies. Chapter 2 describes the Solar System and main characteristics of the major planetary bodies treated in this book. Chapter 3 gives brief descriptions of interior structure, climate, geology and tectonics of Earth, Moon, Venus, Mercury and Mars. Chapter 4 is concerned with rocks, discontinuities, weathering and physico-mechanical properties of rocks and discontinuities with an emphasis on

Mars, and Chapter 5 is concerned with constitutive laws for rocks and discontinuities, which are necessary for not only Earth but also for other terrestrial planetary bodies. The available studies on the stress states on Mercury, Venus, Earth, the Moon and Mars are described, and detailed analyses on their stress states are presented in Chapter 6. In Chapter 7, it is shown that quakes also occur on planetary bodies and that marsquakes and moonquakes do occur. Furthermore, the possibilities of venusquakes and mercuryquakes are discussed. This issue may be quite important regarding the settlement and exploitations of these planetary bodies. The dynamics of impacts and crater formation, together with recent observations, are presented in Chapter 8, which are very important for the damaging seismic effects of impacts on the environment. While Chapter 9 is concerned with slope stability issues on planetary bodies, Chapter 10 explains the stability issues of underground cavities and sinkhole issues on Earth and other terrestrial bodies. Chapter 11 describes hydro-magnesite formation in Salda Lake in Türkiye as an example on Earth in relation to the early life traces in the Jezero Crater of Mars. Chapter 12 is concerned with comets and asteroids, and their features are presented and discussed. Important issues to be dealt with in planetary rock mechanics and rock engineering are presented and discussed in Chapter 13. Finally, the settlement and exploitation aspects in terrestrial planetary bodies are presented, and possible long-term environmental issues are discussed in Chapter 14.

Acknowledgments

The author sincerely acknowledges Prof. Reşat Ulusay and Prof. Xia-Ting Feng, former Presidents of ISRM, and Prof. Seokwon Jeon, the current President of ISRM, for inviting the author to contribute to the prestigious ISRM Book Series on Planetary Rock Mechanics. The author also sincerely acknowledges NASA and the people involved in the development and operation of Mars exploration programs, Mars rovers, namely Opportunity, Spirit, Curiosity and Perseverance, and processing and releasing their images on related websites.

A motivation for the author to become involved in the topic of this book was related to his participation in a special session within the Mining Engineering Conference in Montreal in 2015 and the invitations by Prof. Serkan Saydam to give talks on "Some Thoughts about Rock Mechanics Aspects of Mars" at the 3rd Off Earth Mining Forum—2017OEMF in 2017 and ISRM workshop in 2022 held in the University of New South Wales (UNSW), Sydney, Australia, which made the author to become really involved with Planetary Rock Mechanics.

The images from recent Mars exploration rovers (MERs) by NASA showed striking similarities between rocks on Earth and those of Mars and motivated the author to bring together some of his thoughts about the aspects of rock mechanics and rock engineering in Mars and to compare them with those of Earth in addition to those from other terrestrial bodies. Most of images related to Mars are from Mars Exploration Rovers of NASA and they are gratefully acknowledged. The author also sincerely acknowledges NASA and the people involved in the development and operation of Mars exploration programs, Mars rovers, namely Opportunity, Spirit, Curiosity and Perseverance, and processing and releasing their images on related websites. This book is based on information gained from images mostly released by NASA and the interpretation of the author, together with some further information and interpretations by others.

Finally, I want to thank my wife, Reiko, and my daughter, Ay, and my son, Turan Miray, and my late parents for their continuous help and understanding at all times, without which this book could not have been completed.

Author Biography

Ömer Aydan was born in 1955 and studied Mining Engineering at the İstanbul Technical University, Turkey (BSc, 1979), Rock Mechanics and Excavation Engineering at the University of Newcastle upon Tyne, UK (MSc, 1982), and finally received his PhD in Geotechnical Engineering from Nagoya University, Japan, in 1989. He worked at Nagoya University as a research associate (1987–1991) and at the Department of Marine Civil Engineering at Tokai University, first as Assistant Professor (1991–1993), then as Associate Professor (1993–2001), and finally as Professor (2001–2010). He then became Professor of the Institute of Oceanic Research and Development at Tokai University, and he moved to the Department of Civil Engineering, the University of Ryukyus (Nishihara, Okinawa, Japan) as the professor of Material Science and retired at the end of March 2020 as Emeritus Professor. He has been an adjunct staff of the University of the Ryukyus Disaster Prevention Research Center for Island Regions since 2014 and served as the director of the Center until he retired. He established the natural disaster simulation facility at the Center. Dr. Ömer Aydan has played an active role on numerous ISRM, JSCE, JGS, SRI and Rock Mech. National Group of Japan committees and has organized several national and international symposia and conferences. He organized the 2019 Rock Dynamics Summit in Okinawa in May 7–11, 2019, the world-class conference in Rock Dynamics. He was also made Honorary Professor in Earth Science by Pamukkale University in 2008. He served as the Vice President at Large of the International Society for Rock Mechanics and Rock Engineering for the term of 2019–2023. He was appointed as associate editor of the *Journal of Rock Mechanics and Rock Engineering* in April 2020. He is the author and/or one of the authors of more than 900 scientific papers and the author of eight international books on rock mechanics and has co-authored five other international books.

Introduction

1

1.1 INTRODUCTION

Mankind has been exploring space objects since ancient times. For example, Sumerians, who established a great civilization in Mesopotamia between 4000 and 2000 BCE, laid out the foundations of the modern astronomy. Indian, Chinese, Persian, Arabian and Turkish astronomers fundamentally are similar to those of Sumer. Figure 1.1 shows the symbols of the universe according to Sumerians and Oz Tamga (symbol) of proto-Turks on a basaltic rock surface at Saymalıtaş (Kyrgyzstan) (e.g., E. Herzfeld 1930; Çığ 2020; Somuncuoğlu 2010, 2011). One can find such observatories even today in the cities of Turkistan, such as Semerkand (Ulug Bey Observatory) and Maragheh built during the Turk-Mogol Ilhan Khanate to observe celestial objects and their positions (Heather 1999). Most of these observatories were utilized for creating calendars for agricultural and religious purposes.

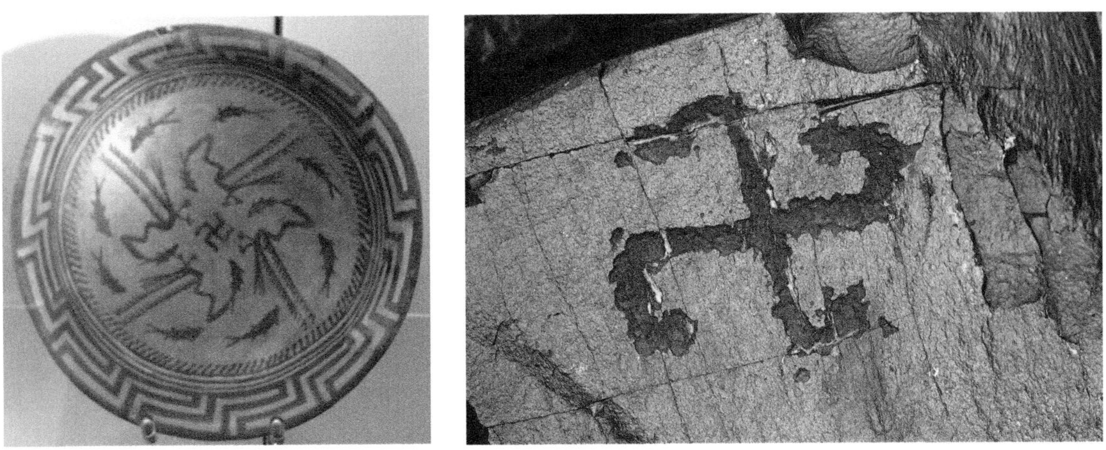

(a) Sumerian bowl symbolizing universe (4000 BCE)

(b) Oz Tamga (symbol) (Saymalıtaş, older than 5000 BCE, likely to be 10000 BCE)

FIGURE 1.1 Artifacts symbolizing the universe from Sumers and proto-Turks.

The US and Russia (former Union of Soviet Socialist Republic (USSR)) initiated space programs and sent several spacecraft to the Moon, Venus and Mars, and some of them landed to find out the characteristics of other planets and possibility of exploiting their mineral resources (NASA 1997, 2008). Some of most impressive images from the Apollo Program of NASA were the astronaut from Earth standing next to a lunar rock mass and a geologist from Earth hitting lunar rock with a geologist's hammer (Figure 1.2). The images from recent Mars exploration rovers such as Opportunity, Spirit,

DOI: 10.1201/9781003643852-1

Curiosity and Perseverance (NASA 2019, 2022) showed striking similarities between rocks on Earth and those of Mars. One can easily notice layered rock masses, jointing, weathering effects and some toppling-type rock slope stability problems (Figure 1.3), as well as sinkholes (skylights) related to the collapse of underground openings.

(a) Broken rock mass next to humankind (b) Geologist hitting lunar rock with a hammer

FIGURE 1.2 Images from the Moon involving humankind from Earth (images by NASA).

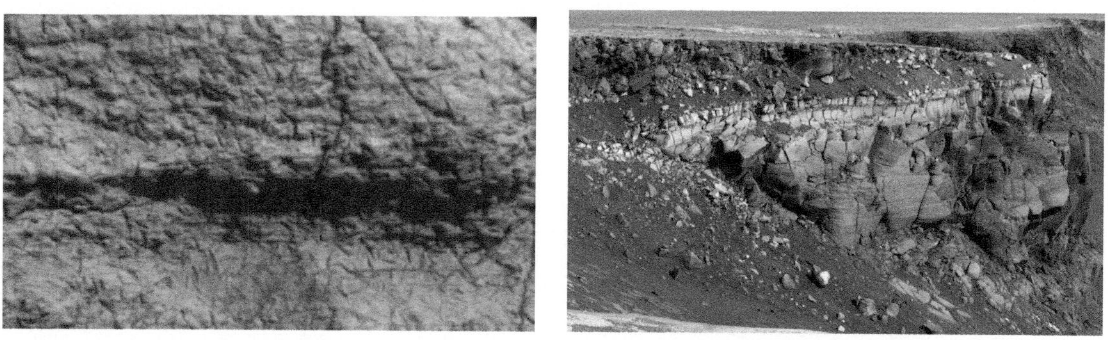

(a) Layered rock mass on Mars (b) Toppling of a rock slope on Mars

FIGURE 1.3 Images of rocks and rock slope on Mars (arranged from images of NASA).

The chapters of this book are based on information obtained from images mostly released by NASA and its associated institutes and interpreted by the author, together with some additional information and publications from the viewpoint of Rock Mechanics and Rock Engineering.

The images from explorations on the Moon and Mars showed the great similarities between rocks on Earth and those on the Moon and Mars, which motivated the author to bring together some of his thoughts about the aspects of rock mechanics and rock engineering in other planetary bodies, specifically Mars and the Moon, and to compare them with those of the Earth.

Although the environmental conditions on Mars, the Moon and other planets are different from those on Earth, the principles governing mechanical and engineering aspects of rocks on other planets are expected to be quite similar to those developed for the rocks of Earth. Therefore, the next generations of our discipline would definitely see its extension to the rocks of other planets. The establishment of the Commission on Planetary Rock Mechanics by the International Society for Rock Mechanics and Rock Engineering (ISRM) in 2019 is one of the first steps to achieve such a goal.

This book covers some thoughts of the author on the aspects of Planetary Rock Mechanics and rock engineering deducted from the images sent to Earth from Mars, the Moon, Mercury and Venus by exploration rovers and from comets and asteroids from spacecraft. The images from Mars and the Moon particularly enable us to imagine and infer many aspects planetary rock mechanics and resemblances of rocks and their weathering, discontinuity formations, slope stability problems and natural caves, as well as possible mechanics and material behavior of rocks and their related scientific and engineering aspects.

The book has 14 chapters. Chapter 1 briefly explains the motivation as well as a summary of planetary bodies. Chapter 2 describes the Solar System and main characteristics of major planetary bodies treated in this book. Chapter 3 gives brief descriptions of interior structure, climate, geology and tectonics of Earth, the Moon, Venus, Mercury and Mars. Chapter 4 is concerned with rocks, discontinuities, weathering and physico-mechanical properties of rocks and discontinuities with an emphasis on Mars, while Chapter 5 is concerned with constitutive laws for rocks and discontinuities, which are necessary for not only Earth and but for also other terrestrial planetary bodies. The available studies on the stress states in Mercury, Venus, Earth, Moon and Mars are described, and detailed analyses on their stress states are presented in Chapter 6. It is shown in Chapter 7 that quakes also occur in planetary bodies and that Marsquakes and Moonquakes do occur, as anticipated. Furthermore, the possibilities of Venusquakes and Mercuryquakes are discussed. This issue may be quite important in the settlement and exploitations of the planetary bodies. The dynamics of impacts and crater formation with recent observations are presented in Chapter 8, which are very important for the damaging and seismic effects of impacts on the environment. While Chapter 9 is concerned with slope stability issues in planetary bodies, Chapter 10 explains the stability issues of underground cavities and sinkhole (skylight) issues. Chapter 11 describes the hydro-magnesite formation in Salda Lake in Türkiye as an example on Earth in relation to the early life traces in the Jezero Crater of Mars. Chapter 12 is briefly concerned with comets and asteroids, and their features are presented and discussed. Important issues to be dealt in planetary rock mechanics are presented and discussed in Chapter 13. Finally, the settlement and exploitation aspects in terrestrial planetary bodies are presented, and possible long-term environmental issues are discussed in Chapter 14.

Solar System and Terrestrial Planetary Bodies

2

2.1 PLANETARY BODIES OF OUR SYSTEM

As is well-known, the Solar System consists of eight planets, namely, Mercury, Venus, Earth, Mars, Saturn, Jupiter, Uranus and Neptune (Figure 2.1). On the basis of computations, the existence of a planet between Mars and Jupiter has been estimated. However, at the estimated location, an asteroid belt exists (Figure 2.2). This asteroid belt might have corresponded to the missing planet. Initially, Pluto was found in 1930, and it was considered to be the ninth planet until 2006. On the basis of the definition of planets in our Solar System, it lost that designation and was not a planet any more. The density of planets differs as a function of distance from the Sun, and it decreases as the distance increases. Terrestrial planets such as Mercury, Venus, Earth and Mars are considered to be quite similar to one another in many aspects, while they are different in terms of size and geology. Mercury and Venus have no natural satellites, whereas the Earth has the Moon as a natural satellite. Mars has two natural satellites called Deimos and Phobos.

FIGURE 2.1 Comparison of Mercury, Venus, Earth and Mars (from NASA).

DOI: 10.1201/9781003643852-2

FIGURE 2.2 Illustration of the Solar System (from NASA).

2.1.1 Earth

The Earth is about 150 million km away from the Sun, and it is 12760 km in diameter. The average density is 5.51 g/cm^3 with a gravitational acceleration of 9.81 m/s^2 at the surface. The Earth completes one rotation around the Sun every 23.9 hours, and it takes 365.25 days to complete one trip around the Sun. Earth's axis of rotation is tilted 23.4 degrees with respect to the plane of Earth's orbit around the Sun. This tilt results in the yearly cycle of seasons.

Earth is rotating around the Sun in a given orbit, and this orbit has an elliptical shape so that the distance between Earth and Sun varies during a year. As a result, the acceleration and velocity of Earth are not the same throughout a year, as shown in Figure 2.3. The stress acting in various parts of Earth would be changing with different periods such as 3 months, 28 days, 24 hours and 6 hours as a result of the motion of Earth around the Sun, daily rotation, rotation of the Moon around Earth due to other cycles, such as the wobbling of the rotation axis of Earth.

Earth is composed of an inner core, enveloped by an outer core, mantle, and crust. The inner core is a solid sphere and is made of iron and nickel metals about 1221 km in radius. The outer core with a thickness of 2300 km surrounds the inner core, and it is made of iron and nickel in the fluid state. The mantle between the outer core and crust has a thickness of 2900 km and is a hot, viscous mixture of molten rock. The outermost layer is Earth's crust that is 30 km thick beneath the continents and about 5 km beneath the oceans.

Earth has volcanoes, mountains and valleys, and its lithosphere consists of the crust and the upper mantle. The lithosphere is divided into huge plates that are constantly moving and causing earthquakes from time to time.

Earth has an atmosphere consisting of 78% nitrogen, 21% oxygen and 1% other gases. The atmosphere shields the planet from much of the harmful radiation coming from the Sun and meteoroids. The average surface temperature on Earth is approximately 14°C, with great variation depending

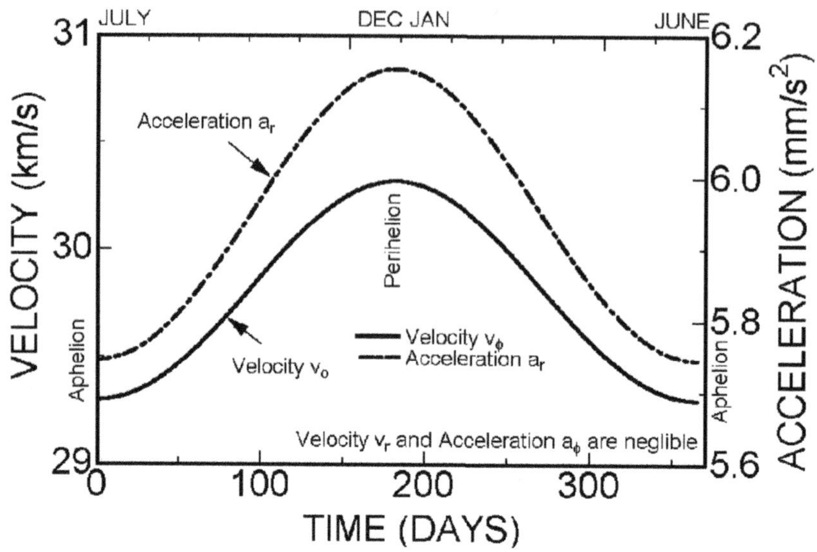

FIGURE 2.3 Acceleration and velocity variations of Earth during a year.

geographical region, elevation and time of year. The highest recorded temperature on Earth is 56.7°C in Death Valley, California, while the coldest temperature is recorded −89.2°C at Vostok Station in Antarctica.

2.1.2 Moon

The Moon is about 284400 km away from Earth, and it is 3474.2 km in diameter. The average density is 3.34 g/cm^3 with a gravitational acceleration of 1.62 m/s^2 at the surface, which is about 1/6 that of Earth. The Moon was assumed to be initially a part of Earth. The Moon is the most extensively investigated solar body by the US and Russia (formerly USSR). Recently Japan, China and India sent unmanned explorers to the Moon. However, the most extensive explorations and investigations have been carried out by NASA of the US so far.

The surface topography of the Moon has been measured by the methods of laser altimetry and stereo image analysis, as shown in Figure 2.4. The surface topography of Moon on the Far side is very different from that on the Near side. The Far side of the Moon is about 1.9 km higher than the Near side. The surface of Moon was greatly shaped by the impact of meteorites. This might imply that the Moon protects Earth from devastating meteoroid impacts. The south pole's Aitken Basin has the lowest elevation, while the highest elevations are found just to the northeast of this basin, and it has been suggested that this area might represent thick ejecta deposits on the Moon. The difference between the lowest and highest elevations is about 13 km. Although there are no tectonic plates on the Moon, huge rift valleys are found, and they are covered with numerous ancient volcanic plains.

2.1.3 Venus

Surface observations of Venus have been difficult in the past, due to its extremely dense atmosphere, which is composed primarily of carbon dioxide with a small amount of nitrogen. The first attempts

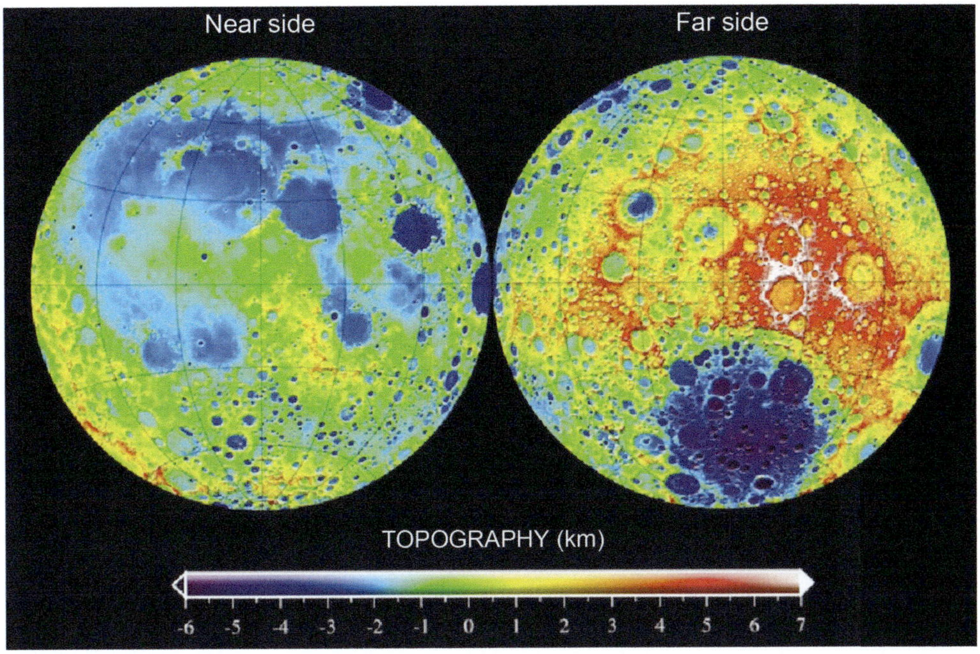

FIGURE 2.4 Surface topography of the Moon (from NASA).

to explore Venus were carried out by the Soviets in the 1960s in the Venera Program (e.g., Sagdeev and Moroz 1982; Rzhiga 1988). Venera lander missions continued until the early 1980s. Venera 7–14 successfully landed on the planet, and Venera 9 was the first to send the first color photographs of the surface. Images taken by Venera 10, 13 and 14 were relayed to Earth. Unfortunately, there are no photographs after the Venera program was terminated. Venera 15 and 16 were the first to conduct mapping of the Venusian terrain with synthetic aperture radar (Rzhiga 1988).

The United States launched Mariner 1 in 1962 and subsequently NASA's Magellan spacecraft in 1989 to map the surface of Venus with radar. The Magellan provided the most high-resolution images to date of the planet and was able to map 98% of the surface and 95% of its gravity field (Figure 2.5).

Venus, like Uranus, rotates clockwise, which is known as a retrograde rotation. Due to the slow rotation on its axis, it takes 243 Earth-days to complete one rotation. The orbit of the planet takes 225 Earth-days.

Venus is totally enshrouded by clouds more than 50 km thick. Due to the continuous global cloud cover and greenhouse effect, surface temperatures of Venus approach 462°C, and it is the hottest planet in our Solar System. As a result of these high temperatures, water is vaporized. The atmospheric pressure at the surface of Venus is 92 times (9.2 MPa) that found at the surface of Earth.

2.1.4 Mars

The radius of Mars is 3389 km, which is almost the half of the radius of Earth, and its gravitation acceleration at ground surface is 0.377 g of Earth. The mean density of Mars is about 3.9335 g/cm^3 on the basis of the law of the universal gravitation. Mars has a very thin atmosphere consisting of carbon dioxides, nitrogen, argon gases, and it is a dusty, cold, desert environment. The rotation axis is tilted

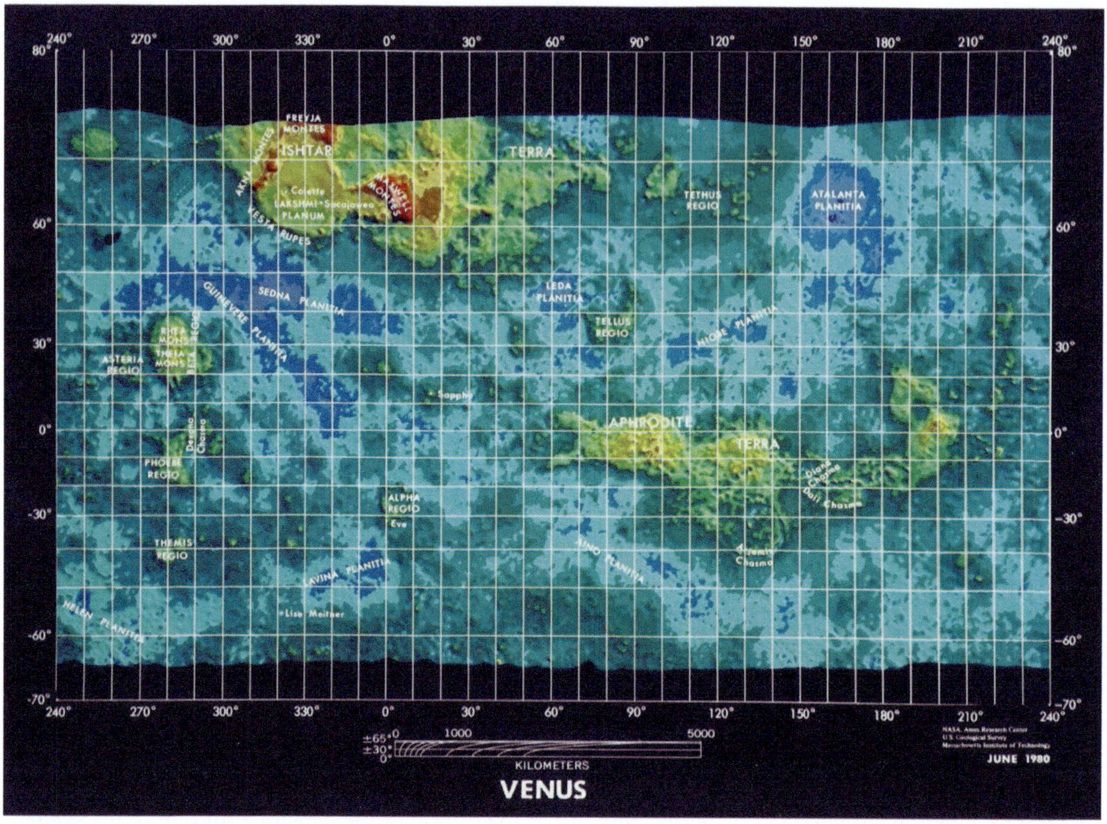

FIGURE 2.5 Topography of Venus (from NASA).

by 25 degrees with respect to the plane orbit around Sun, and the year of Mars lasts about 687 earth days and 669.6 solar days (abbreviated as sols). Furthermore, its rotation about its axis takes about 24.6 hours.

There are four seasons in Mars, although their duration differs. Like Earth, temperature varies depending upon altitude and latitude. The highest temperature may be up to 20°C while the minimum temperature may be minus 153°C. The daily temperature difference at a given location may be more than 80°C (Figure 2.6).

The geography of Mars varies considerably, and the northern plains are flat due to lava flows, in contrast with the southern highlands, which are pitted and cratered by ancient impacts. The surface of Mars as seen from Earth is consequently divided into two kinds of areas (Figure 2.7).

Mars has the highest volcano, Olympus Mons (Mount Olympus), and it rises 22 km above the surrounding volcanic plains. It is in a vast upland region called Tharsis, which contains several large volcanoes. The Tharsis region of Mars also has the Solar System's largest canyon system, Valles Marineris or the Mariner Valley, which is 4000 km long and 7 km deep. Mars has countless impact craters.

Mars has two permanent ice caps at the north and south poles. The difference between Mars's highest and lowest points is nearly 30 km (from the top of Olympus Mons at an altitude of 21.2 km to Badwater Crater.

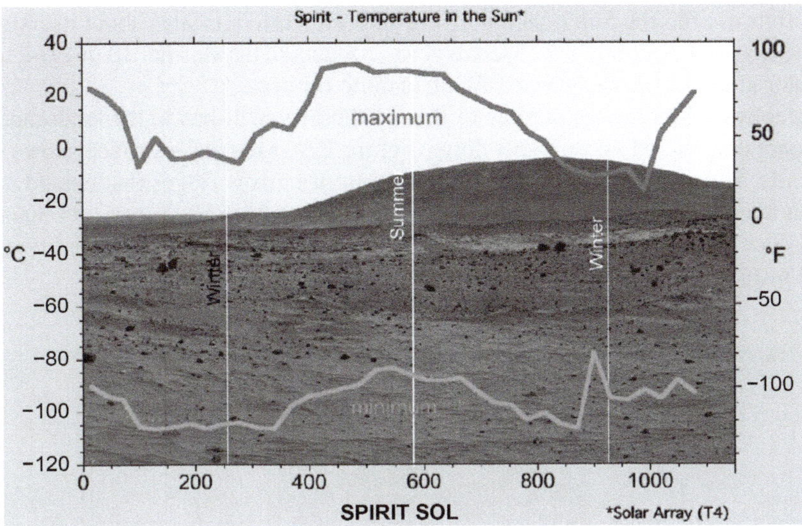

FIGURE 2.6 Yearlong temperature variation in Mars (NASA, Spirit Rover).

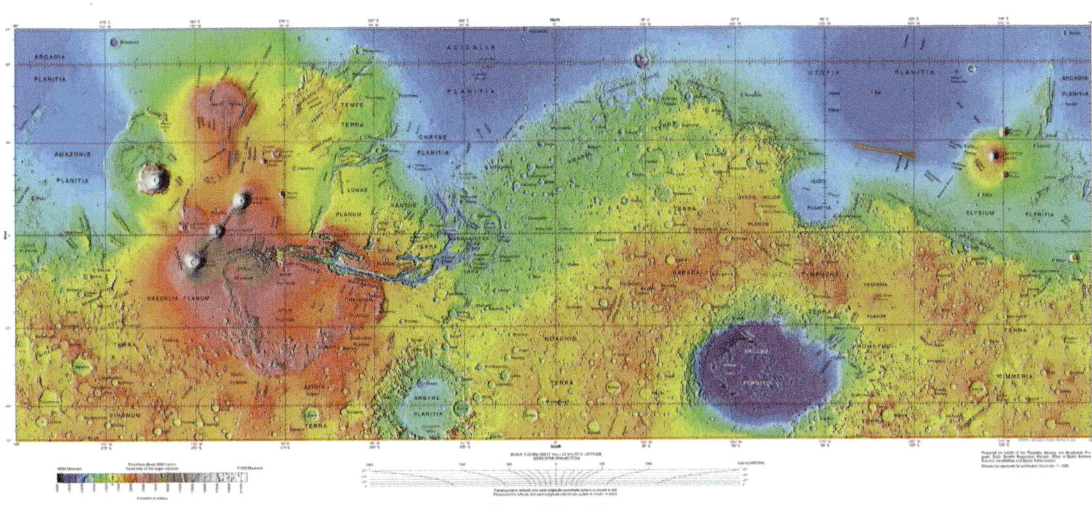

FIGURE 2.7 Planet Mars—topographical map (from NASA).

2.1.5 Mercury

Mercury is the smallest planet in our Solar System with a diameter of 4880 km and the nearest to the Sun. Only slightly larger than Earth's Moon, Mercury formed about 4.5 billion years ago. Like other terrestrial planets, Mercury has a central core, a rocky mantle, and a solid crust. Mercury is the second densest planet after Earth, and its average density is 5.427 g/cm^3. The gravity at the Mercury's surface is 3.7 m/s^2.

Mercury orbits around the Sun in about 88 earth days, while it rotates about its axis about 58.646 Earth-days. The first successful flyby of Mercury was conducted by Mariner 10 in 1974, and it has been visited and explored by the Messenger and BeniColombo orbiters.

The topography of Mercury is similar to that of the Moon. It has a dry landscape pockmarked by asteroid impact craters and ancient lava flows (Figure 2.8). Mercury's surface shows many signs of being geologically active in the past, mainly in the form of narrow ridges that extend up to hundreds of kilometers in length. It is believed that these were formed as Mercury's core and mantle cooled and contracted at a time when the crust had already solidified. However, it is presumed that its geological activity ceased billions of years ago and that its crust has been solid ever since.

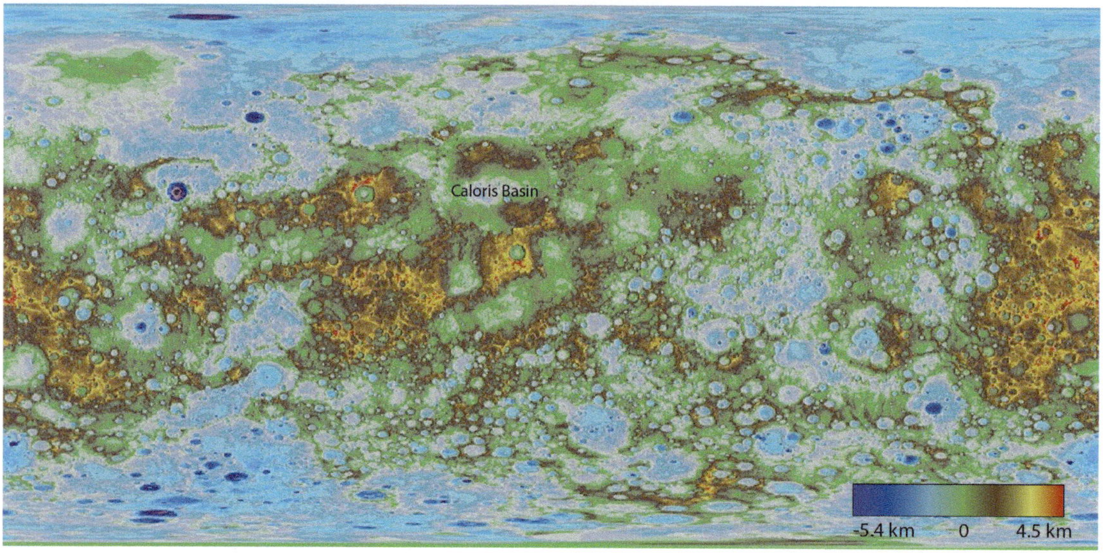

FIGURE 2.8 Planet Mercury—topographical map (from NASA).

Mercury possesses a thin exosphere made up of atoms blasted off the surface by the solar wind and striking meteoroids. Mercury's exosphere is composed mostly of oxygen, sodium, hydrogen, helium and potassium, and it is not able to retain the heat from the Sun. As a result of this and its high eccentricity, the planet experiences far greater extreme variations in temperature than Earth does. Whereas the side that faces the Sun can reach temperatures of up to 427°C, the side that is in darkness can reach temperatures as low as −173°C. Despite these highs in temperature, the existence of water ice and even organic molecules has been confirmed on Mercury's surface. Similar to Earth, Mercury has the presence of water ice in its polar regions.

Interior Structure, Geology and Tectonics of Terrestrial Planetary Bodies

3

3.1 EARTH

This section treats Earth as a planetary body, and it is concerned with its interior structure, geology and tectonics. The evaluations for Earth are extended to other planetary bodies.

3.1.1 Interior Structure

The investigation of the interior structure of Earth requires the sampling of constituting materials and physical, mechanical, chemical experiments. Although boring is used as a technique for sampling, it is limited to shallow depths, even though there have been attempts to drill Earth to a depth up to 40 km. Such an attempt was done in the Kola Peninsula in the former USSR. Similar attempts have been in the US, Germany and recently in China. The cost and difficulties during drilling operations are big handicaps, and such studies are quite limited.

As an alternative tool of investigations, waves travelling through Earth have been utilized. Waves can be induced by seismic events, blasting or other means. Particularly, the waves induced by earthquakes are widely utilized since the advent of seismometers (e.g., Milne 1898).

3.1.1.1 Earthquake-Induced Waves

It is known that earthquakes cause fundamentally two types of waves (Figure 3.1). The first type of waves are called body waves, and they are called P-waves and S-waves. P wave, or primary waves, pass through all material. S-waves, or secondary waves, arrive at the observation point after the P-waves. The second type of waves are called surface waves, and they are further subdivided into Rayleigh- and Love-waves. Figure 3.2 shows a record of the1939 Erzincan earthquake taken at Harvard University. It

DOI: 10.1201/9781003643852-3

is known that shear waves are not transmitted through materials in the liquid phase. Surface waves (Rayleigh and Love) are observed near the earth surface, and they disappear as the depth increases.

The equation of motion governing the propagation of waves irrespective of the state of the given materials can be rewritten in the following form (e.g., Eringen 1980; Mase 1970; Aydan 2017a, 2017b, 2021c).

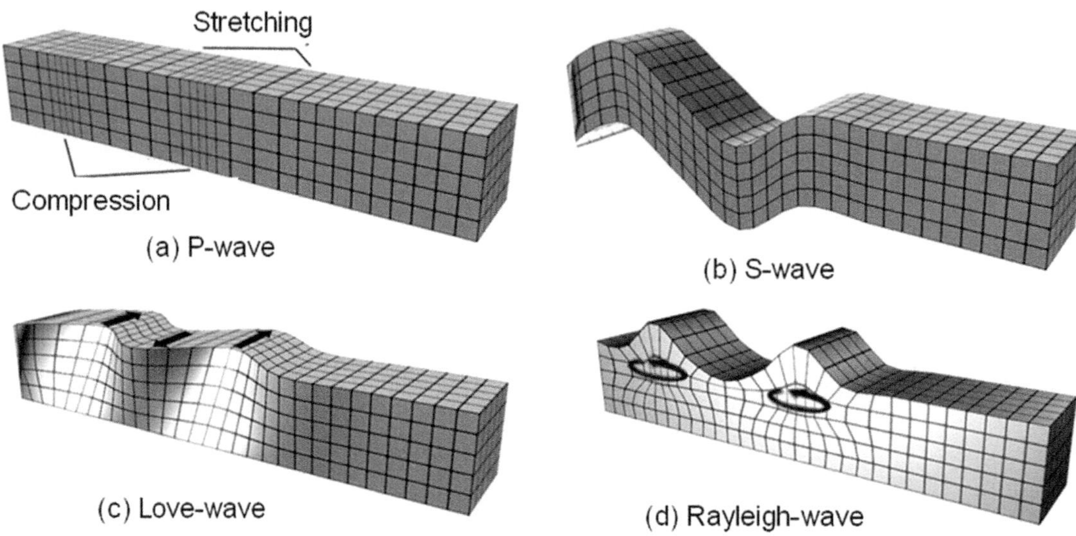

FIGURE 3.1 Illustration of wave types (Aydan 2023a).

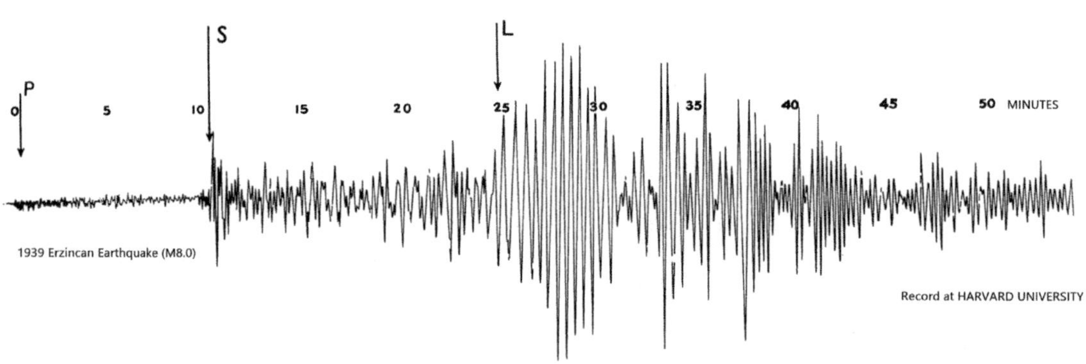

FIGURE 3.2 Seismogram of the 1939 Erzincan earthquake at Harvard University (Ketin 1973).

$$\frac{\partial \sigma_{ij}}{\partial x_j} + b_i = \rho \frac{\partial^2 u_i}{\partial t^2} \tag{3.1}$$

If the material behavior is assumed to be elastic, Eq. (3.1) can be converted to the following form:

$$(\lambda + \mu)\nabla^2\Delta + \mu\nabla^2\Delta = \rho\frac{\partial^2 \Delta}{\partial t^2} \text{ or } V_p^2\nabla^2\Delta = \frac{\partial^2 \Delta}{\partial t^2} \text{ or } V_p{}^2\nabla^2\varepsilon_v = \frac{\partial^2 \varepsilon_v}{\partial t^2} \tag{3.2}$$

where

$$\lambda + 2\mu = \frac{Ev}{(1+v)(1-2v)} + \frac{E}{1+v} = \frac{E(1-v)}{(1+v)(1-2v)}$$

$$V_p = \sqrt{\frac{E(1-v)}{\rho(1+v)(1-2v)}},$$

$$\varepsilon_v = \Delta$$

where E is the elastic modulus, v is Poisson's ratio.

Eq. (3.2) is known as the governing equation of P-wave propagation in solids. As noted from this equation, P-wave propagation is directly related to volumetric straining. During the propagation of the P-wave, the solid undergoes dilatational and compressive volumetric straining.

Using the rotational strain definition and division by density yields the following equations:

$$\frac{\mu}{\rho}\nabla^2\omega_1 = \frac{\partial^2\omega_1}{\partial t^2};\ \frac{\mu}{\rho}\nabla^2\omega_2 = \frac{\partial^2\omega_2}{\partial t^2};\ \frac{\mu}{\rho}\nabla^2\omega_3 = \frac{\partial^2\omega_3}{\partial t^2} \tag{3.3}$$

Eq. (3.3) is the governing equation of Rayleigh-waves. The coefficient in Eq. (3.3) is interpreted as the propagation velocity of rotational waves.

$$V_s = \sqrt{\frac{\mu}{\rho}} \text{ or } V_s = \sqrt{\frac{E}{2\rho(1+v)}} \tag{3.4}$$

If volumetric strain (Δ) is nil and body force is negligible, one easily gets the following expressions:

$$\frac{\mu}{\rho}\nabla^2 u_1 = \frac{\partial^2 u_1}{\partial t^2} \text{ or } V_s^2\nabla^2 u_1 = \frac{\partial^2 u_1}{\partial t^2} \tag{3.5a}$$

$$\frac{\mu}{\rho}\nabla^2 u_2 = \frac{\partial^2 u_2}{\partial t^2} \text{ or } V_s\nabla^2 u_2 = \frac{\partial^2 u_2}{\partial t^2} \tag{3.5b}$$

$$\frac{\mu}{\rho}\nabla^2 u_3 = \frac{\partial^2 u_3}{\partial t^2} \text{ or } V_s\nabla^2 u_2 = \frac{\partial^2 u_2}{\partial t^2} \tag{3.5c}$$

Eq. (3.5) is the fundamental equation of distortion (shear) waves known as S-waves. It should be noted that the propagation velocity of S-waves is the same as that of Rayleigh-waves.

3.1.1.2 Wave Propagation through Earth and Inference of Earth Interior

As said previously, waves are categorized into body and surface waves induced by earthquakes, which travel throughout the Earth. The propagation of surface waves is restricted to the observations near the surface of Earth. P-waves, or dilatational waves, can propagate any medium, while shear waves cannot propagate through fluids. This feature of waves is very important for the interpretation of wave observations. In a uniform homogeneous medium, the wave ray would be straight line, while the variation of density and elastic constants of medium would change the ray path, and the abrupt change of medium would cause refraction. Utilizing the wave propagation theory and Snell's law, the analyses of records at seismographs installed at various locations on Earth revealed the interior of Earth as illustrated in Figure 3.3. If Earth is considered as a simple three-layered structure consisting of mantle crust, outer core and inner core with different densities and wave velocity characteristics and outer core is fluid, there will be zones of shadow for P-waves and S-waves. The P-wave shadow occurs between 103 and143 degrees due to the refraction caused by the characteristics of mantle crust and outer core, while the S-wave shadow is due to no S-wave propagation through the outer core. As the outer core of Earth is in liquid phase, no shear waves are observed beyond 103 degrees from the earthquake focus.

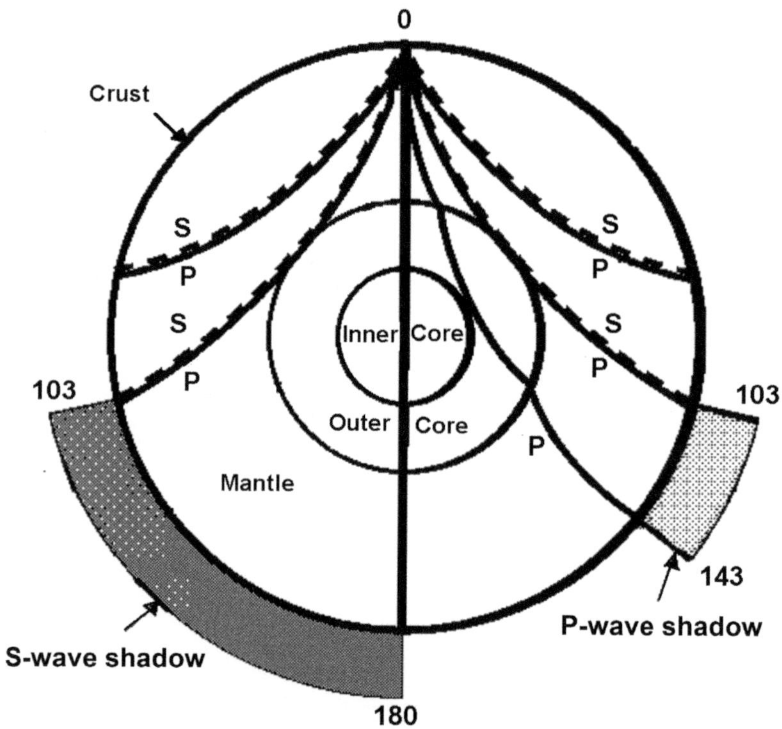

FIGURE 3.3 Wave propagation and interior structure of Earth.

As shown in Figure 3.3, various waves induced by an earthquake would be observed, the waves will continue to appear on the seismic records for a certain period of time, and they will eventually disappear due to volumetric and viscous dispersion. In seismology, the observed waves are distinguished according their path and the media they propagated through, as illustrated in Figure 3.4. For example, P-wave would correspond to the first arrival of P-wave, while PP would correspond to a reflected P-wave. Waves denoted PKP would involve a path through the mantle (P)–outer core (K)–mantle (P), while PKIKP would involve a path through mantle (P)–outer core (K)–inner core (I)–outer core (K)–mantle (P). P-waves can be reflected as S-waves. In such a case, the observed wave would be denoted as PS. Figure 3.5 shows an example of records observed at different locations on Earth. The distance is measured as the angle from the focus of an earthquake.

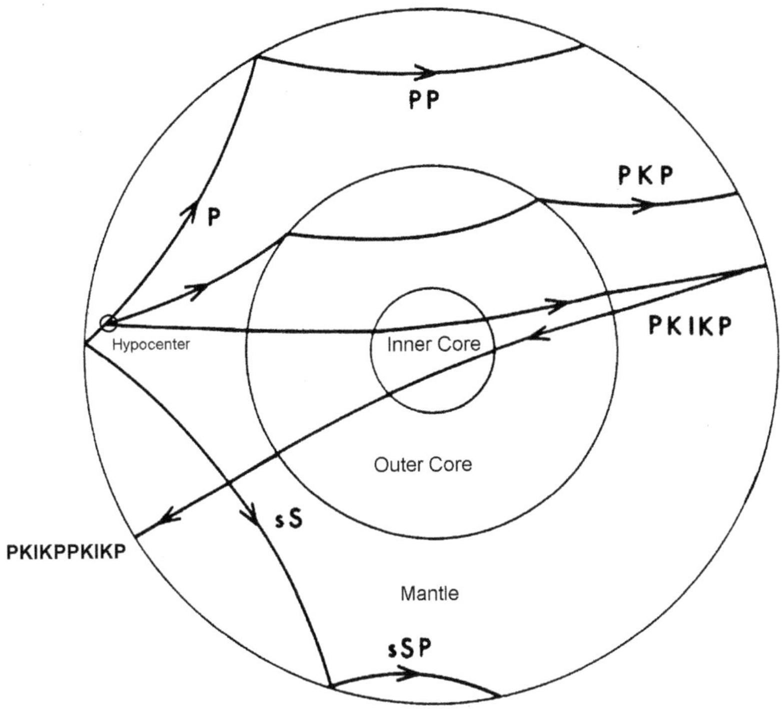

FIGURE 3.4 Illustration of various arrival waves.

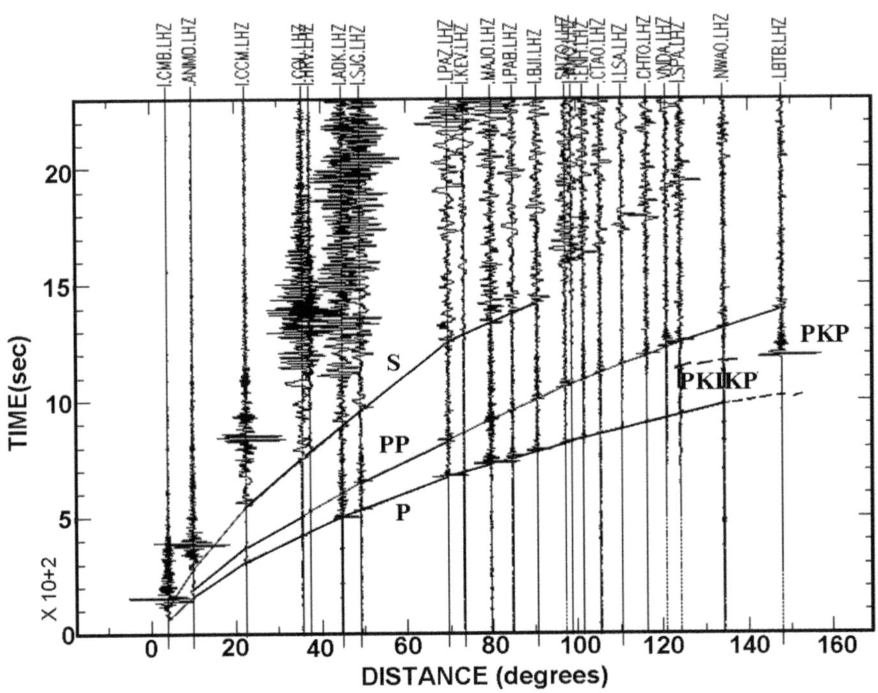

FIGURE 3.5 Observed wave forms and various phase arrivals.

Milne (1898), who was originally a mining engineer from UK, together with T. Gray, was the first to develop the modern age seismographs following the 1880 Yokohama earthquake in Japan. Seismographs were updated continuously since then, and there are many seismic networks, internationally, country-wise and locally, these days. The analyses of seismic records and several techniques developed for interpretation and inferences (e.g. Jeffreys and Bullen 1940), the interior structure of Earth was contemplated as shown in Figure 3.6. Based the density, P-wave and S-wave distributions and some thermo-mechanical tests on rocks and materials existing on Earth, it is shown that Earth broadly consists of core, mantle and crust. While the outer core is assumed to be in the fluid state in view of its shear wave velocity, the inner core is presumed to be solid or solid-like. The mantle is divided into two layers, namely, the upper mantle and the lower mantle. Particularly, the upper mantle seems to be highly disturbed. The crust of the earth varies from place to place. While it is about 33 km thick on average, it becomes quite thick in the close vicinity of subduction zones and thinner near the divergent plate boundaries. The interior structure models of other planets are fundamentally based on this concept.

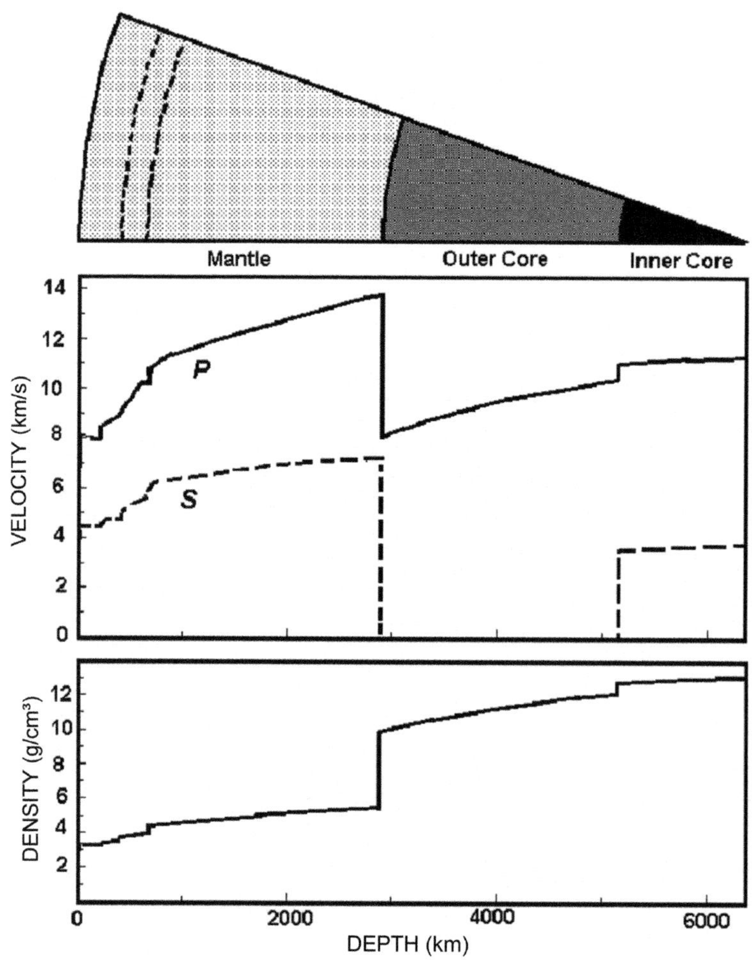

FIGURE 3.6 Inferred interior structure of Earth.

3.1.2 Tectonics

The tectonics of Earth utilizes the plate tectonic theory utilizing Wegener's theory on continental drift proposed in 1912, and it is often used to explain the occurrence of quakes of Earth, called earthquakes. However, Wegener's theory was not well-received among mainstream geologists and geophysicists until the discovery of paleomagnetism and modern seismology. Figure 3.7 shows the plots of earthquake epicenters and the major plates of Earth.

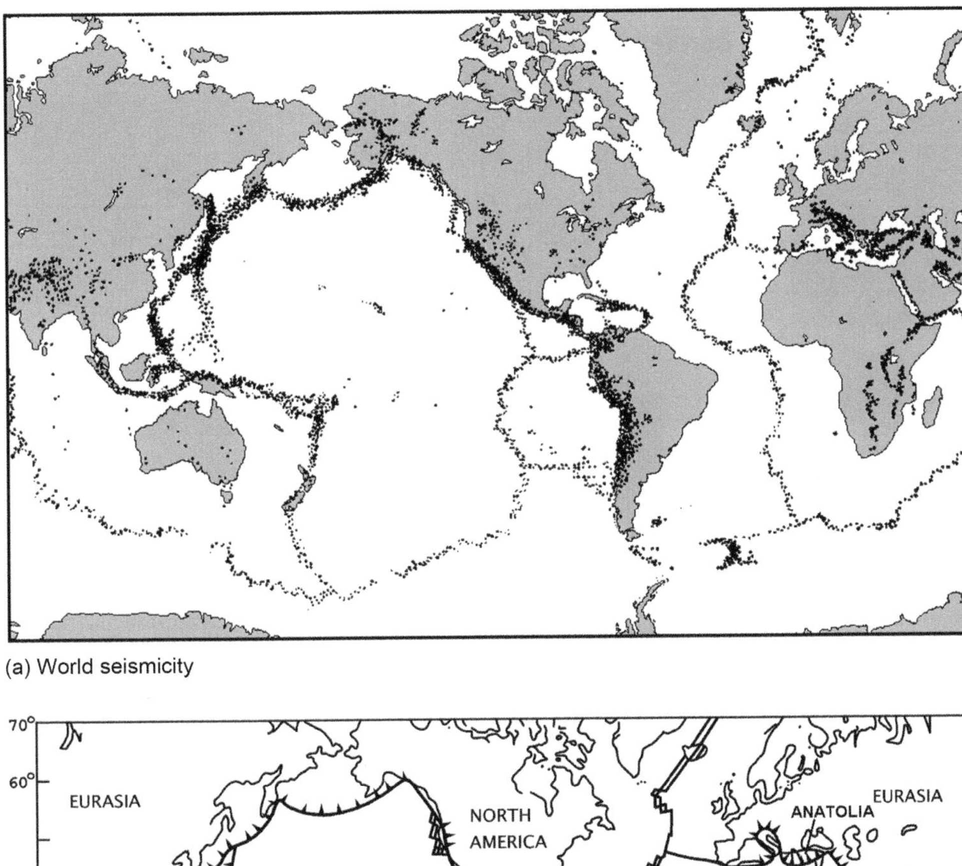

(a) World seismicity

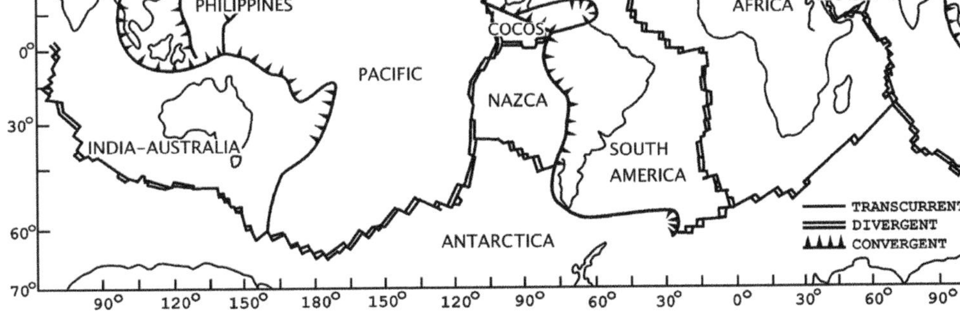

(b) Major plates of Earth (modified from Udias, 1999)

FIGURE 3.7 (a) World seismicity and (b) major plates of Earth.

It is often assumed that seismic regions coincide with plate boundaries. There are three major types of plate boundaries, namely convergent, divergent and transform plate boundaries, as illustrated in Figure 3.8. High mountains are found at/along convergent boundaries and plate thickness increases up to 70 km. Furthermore, large earthquakes occur along convergent plate boundaries while the magnitude of earthquakes along divergent plate boundaries is generally smaller. The transform plate boundaries may have dextral or sinistral sense of relative slip, and the magnitude of earthquakes generally ranges between those of earthquakes caused by convergent and divergent plate boundaries.

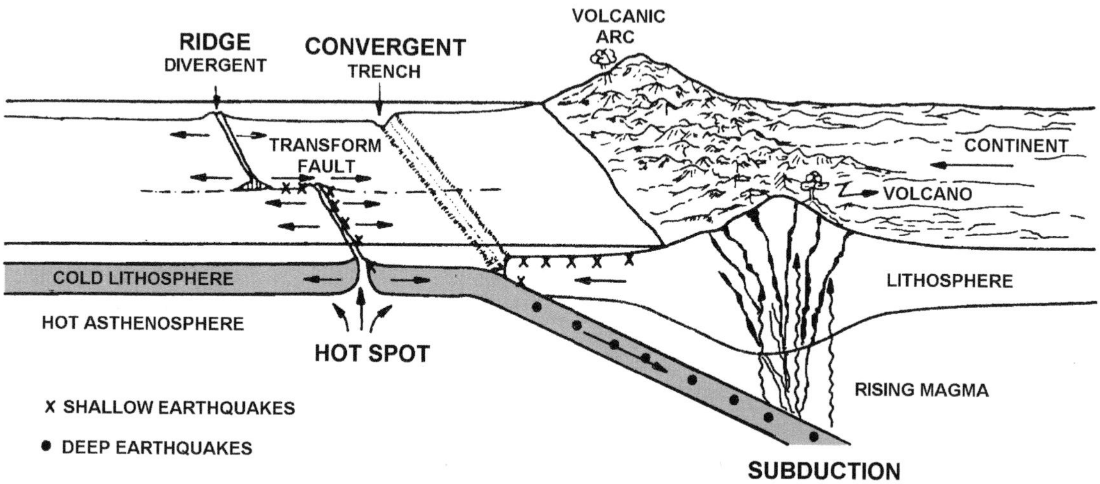

FIGURE 3.8 Types of plate boundaries

The driving force of plates is often said to be due to the mantle convection as illustrated in Figure 3.9. However, it is well known that the convection in Earth cannot take place if Earth consists of uniformly distributed co-centric layers of materials. The subducting plate along the convergent plate boundaries and the divergent plate boundary constitutes the mantle convection cell. This subducting plate is likely to be caused by the plastic yielding of the upper mantle and Earth's crust as explained in the next section.

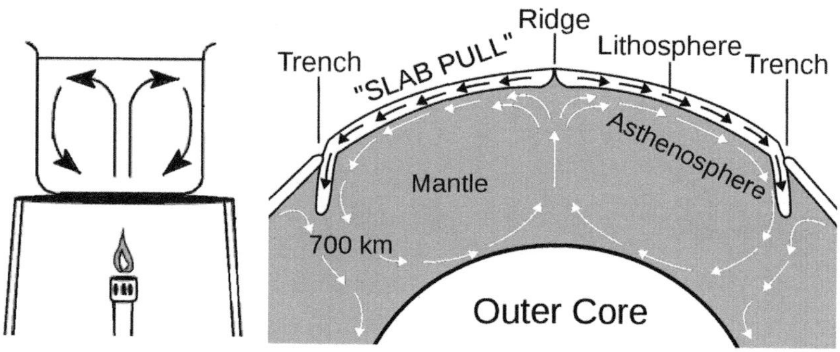

(a) Convection cell (b) Mantle convection cells in Earth

FIGURE 3.9 Simple convection cell and mantle convection cells.

As explained in Chapter 2, Earth is rotating around the Sun in a given orbit, and this orbit has an elliptical shape so that the distance between Earth and the Sun varies during the year. Figure 3.10 shows the effects of the Moon and the Sun on the stress changes of Earth due to their gravitational pull. The stress acting in various parts of Earth would be changing with different periods such as 3 months, 28 days, 24 hours and 6 hours as a result of the motion of Earth around the Sun, daily rotation, rotation of the Moon around Earth, besides other cycles such as the wobbling of the rotation axis of Earth.

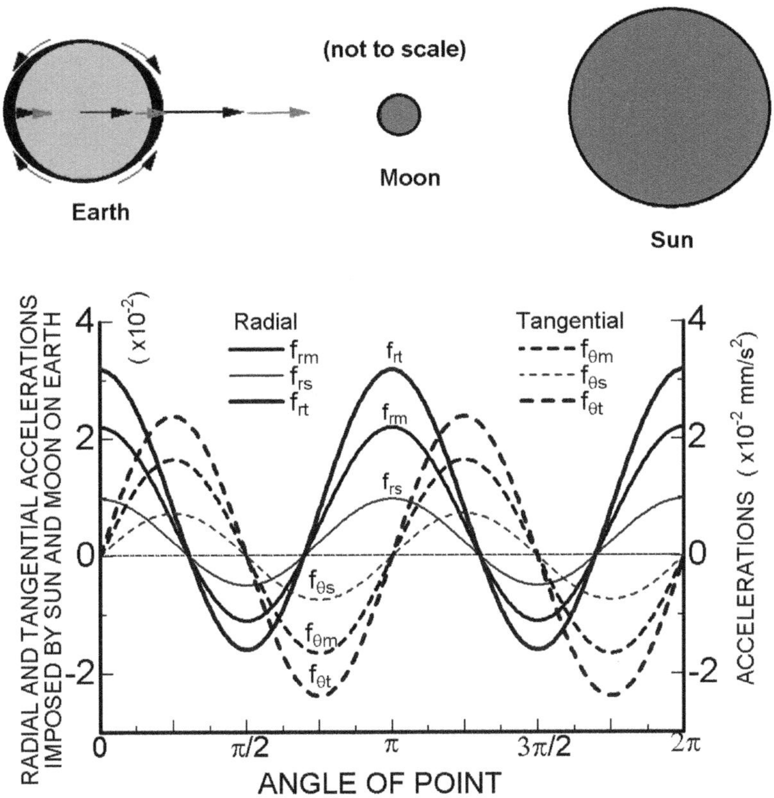

FIGURE 3.10 Effects of the Sun and the Moon on Earth's radial and tangential accelerations.

3.2 THE MOON

3.2.1 Interior Structure

Some passive type seismometers were installed on the Moon in order to understand its interior structure during Apollo program of NASA (1997). During this program from 1969 to 1972, Apollo 12, 14, 15 and 16 seismometers were installed (Figure 3.11). On the basis of these investigations, the inner structure of the Moon is well-understood, and it resembles that of the Earth except for the thickness of each layer (Figure 3.12). Particularly, the outer core is quite thin, and inner core is small.

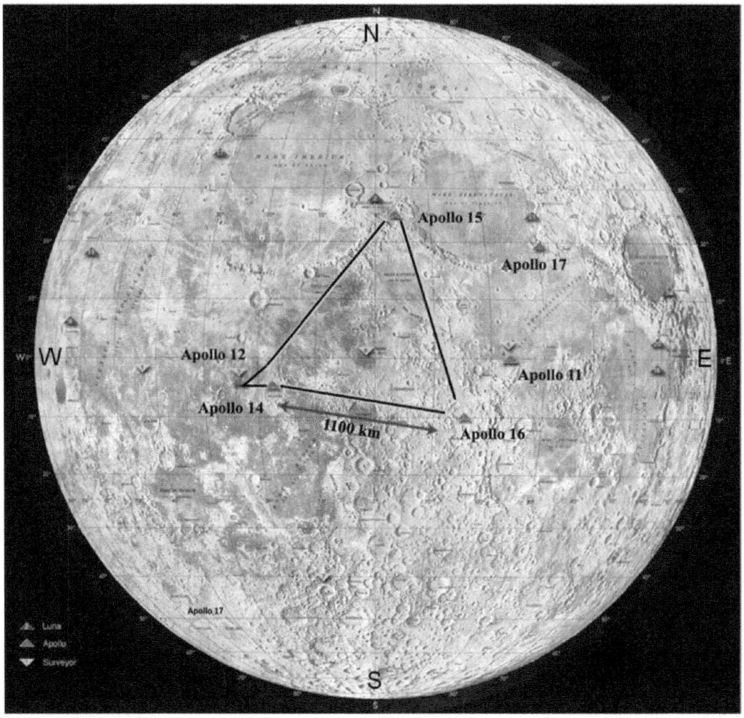

(a) Location of seismometers

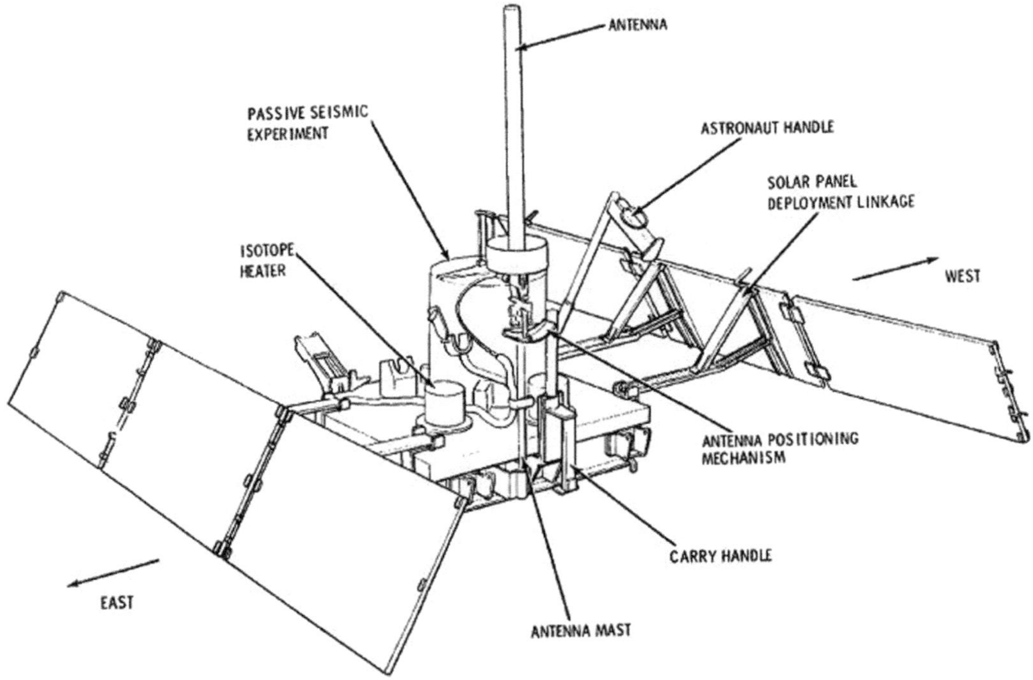

(b) Passive seismometer

FIGURE 3.11 Seismometers and lunar seismic network (from NASA).

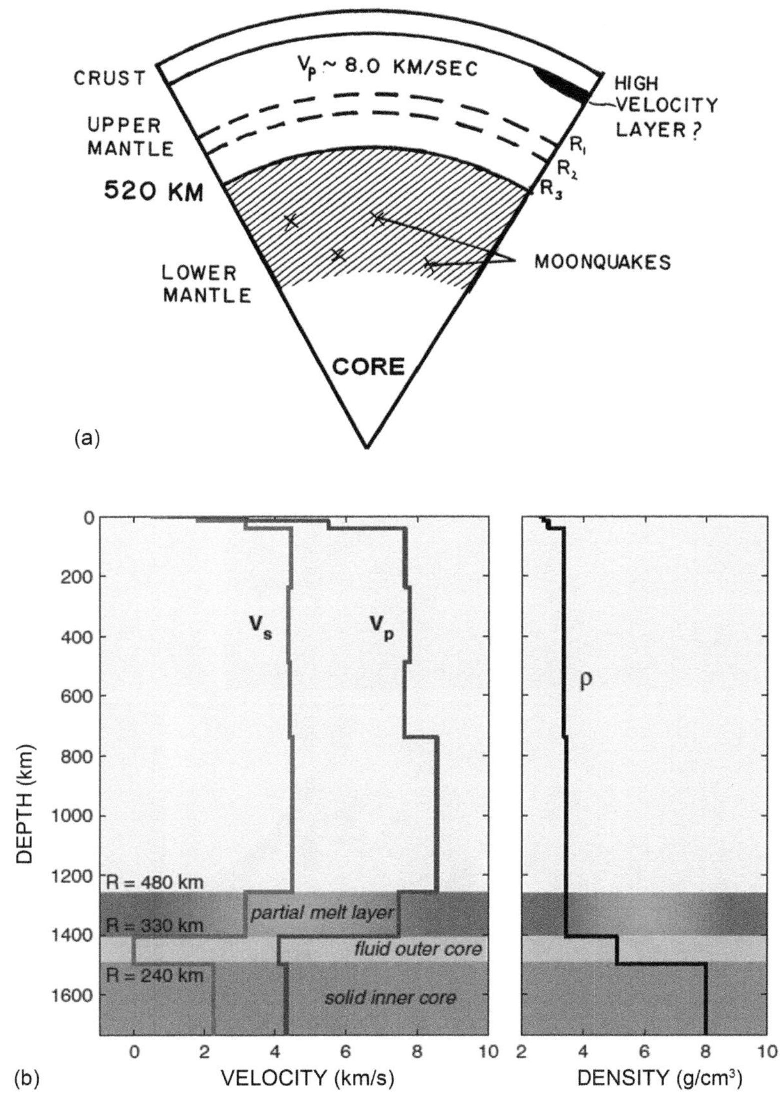

FIGURE 3.12 (a) Inner structure and (b) seismic wave velocity and density distributions.

3.2.2 Geology and Crustal Structures

The geological history of the Moon has been defined into six major epochs starting about 4.5 billion years ago (e.g., Toksöz 1974, 1975; Latham et al. 1971; Nakamura et al. 1974; Weber et al. 2011). It was initially in a molten state. On the basis of the knowledge of Earth, it was assumed that olivine and pyroxene formed first, followed by anorthositic plagioclase feldspar crystallized, forming an anorthositic crust about 50 km in thickness.

The mean lunar surface temperature varies from 107°C at day down to −153°C at night, for a temperature difference of 260°C. This large temperature difference is very likely to create large thermal stresses, which may impose cyclic contraction–expansion as well as new crack extension in the upper crust of the Moon. It is noted that huge thrust- and extension-type cracks/fractures and lava flows exist on the surface of the Moon (Figure 3.13). The active seismic experiments on the Moon during the NASA Apollo program revealed that the top 20 km of the Moon's crust is highly heterogeneous and disturbed.

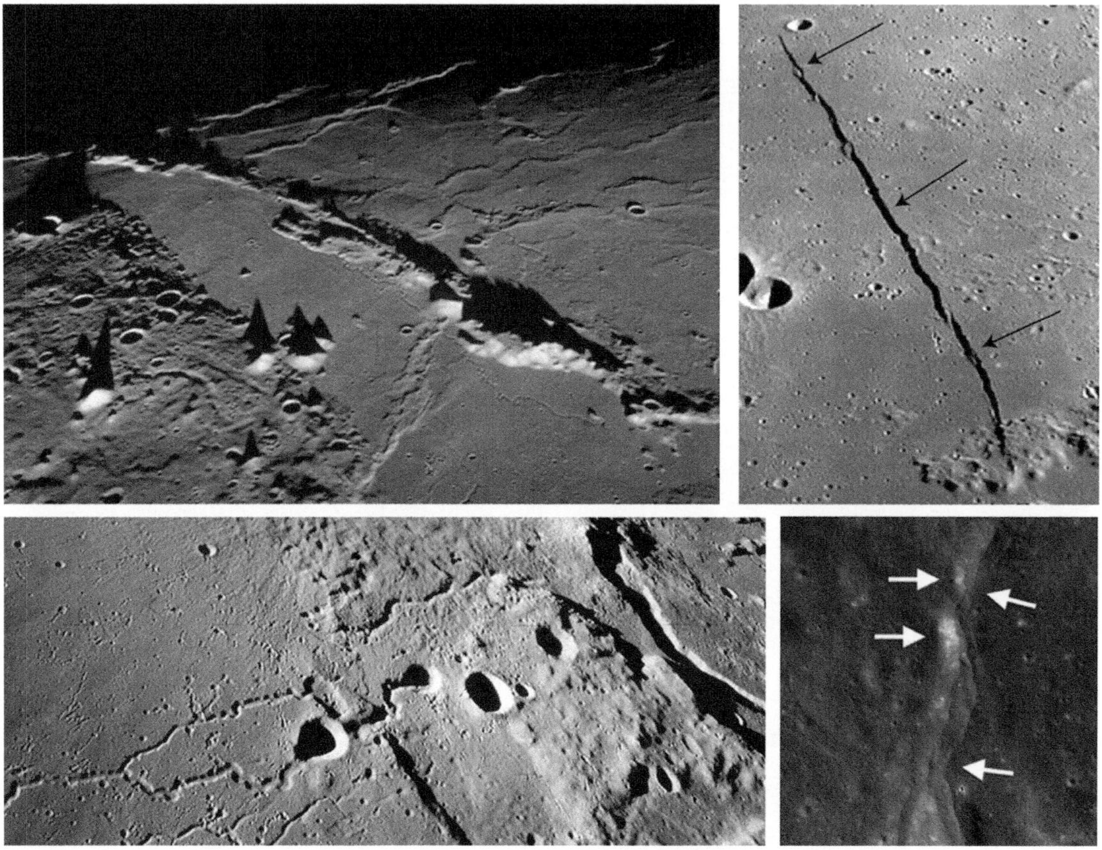

FIGURE 3.13 Large-scale surface fractures on the Moon surface (arranged from NASA).

The Moon has been heavily bombarded by meteorites of different sizes, creating craters (Figure 3.14). The surface topography of the Moon is greatly affected and shaped by the impacts of meteorites. In some large craters, melted rocky material has been noted. In addition, large-scale mass movements exist within the craters.

3.3 VENUS

3.3.1 Interior Structure

It is quite likely that Earth and Venus have similar concentrations of the major elements such as iron, magnesium, calcium, silicon and aluminum. This notion is supported by density measurements

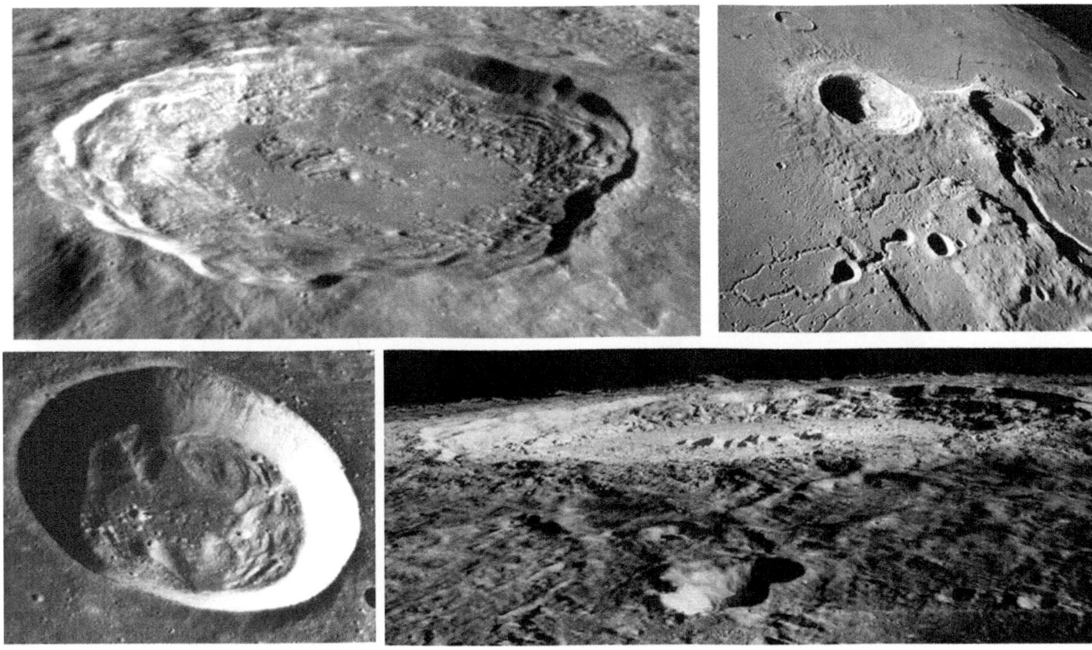

FIGURE 3.14 Craters and large-scale mass movements on the Moon (arranged from NASA).

determined from the motions of passing spacecraft. The bulk density of Venus (5.24 g/cm^3) is only slightly less than that of Earth (5.52 g/cm^3). Venus has a metallic core composed predominantly of iron, a mantle of dense iron and magnesium silicates, and a crust of lighter silicate minerals possibly enriched in aluminum, alkalis and the radioactive elements uranium and thorium.

Similar to the other terrestrial planets, the interior of Venus is essentially composed of three layers: a crust, a mantle, and a core. It is believed that Venus's crust is 8–40 km thick depending upon the assumed model, its mantle 3000 km thick and the core about 3000 km thick (Figure 3.15). However, there is an argument about whether the core of the Venusian interior is liquid or solid, as the planet lacks a substantial magnetic field due to its slow rotation.

Venus is presumed to have emerged about 4.6 billion years ago by the accretion of tiny objects. During the bombardment phase, enough heat was produced to melt the whole proto planet. After a certain cooling period, the molten mass developed a crust, a mantle and a core. Convection in the mantle caused by internal heating deformed the outer crust, which is thinner in low-lying areas (corresponding to seas on Earth) and thicker in the highlands (corresponding to the continents on Earth). High mountain regions developed by uplifting and outflowing.

3.3.2 Tectonics

The tectonics of Venus are quite different from Earth's (Figure 3.16). As the surface temperature of Venus is greater than 460°C, the crustal rocks are likely to be softer and weaker than those of Earth and to behave in a ductile manner. As a result of these features of crustal rocks, the stress built up in the crust and upper mantle of the Venus would be quite smaller than that on Earth. Consequently, the subduction of the soft plastic crust into the upper mantle is not deep. Compared to the subduction of Earth's crust into the upper mantle, which is up to 670 km deep, the subduction of Venus's crust is

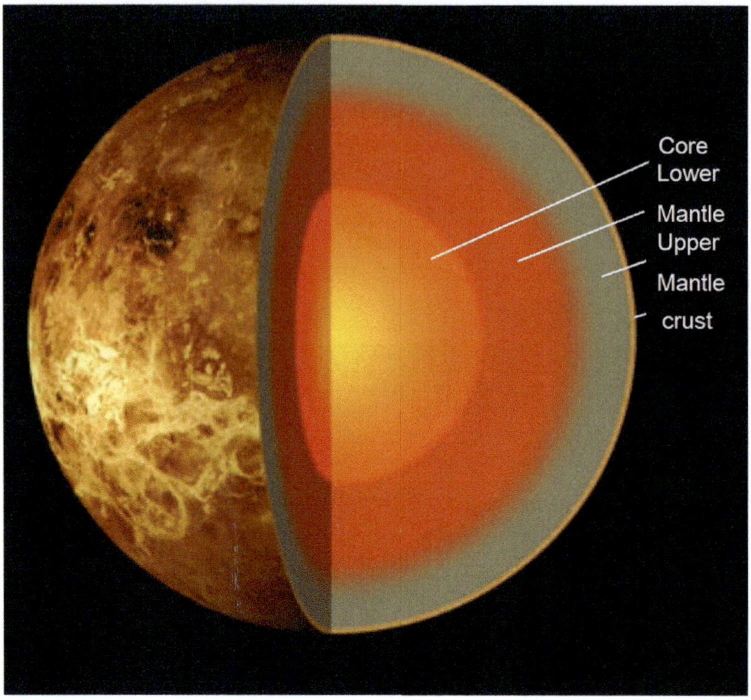

FIGURE 3.15 Cutaway illustration of Venus interior (modified from NASA).

estimated to be 250–300 km deep. In view of these facts, the size and the drag of the mantle convection in Venus should be quite a bit weaker than those of Earth so that the tectonic motions of Venus are less pronounced.

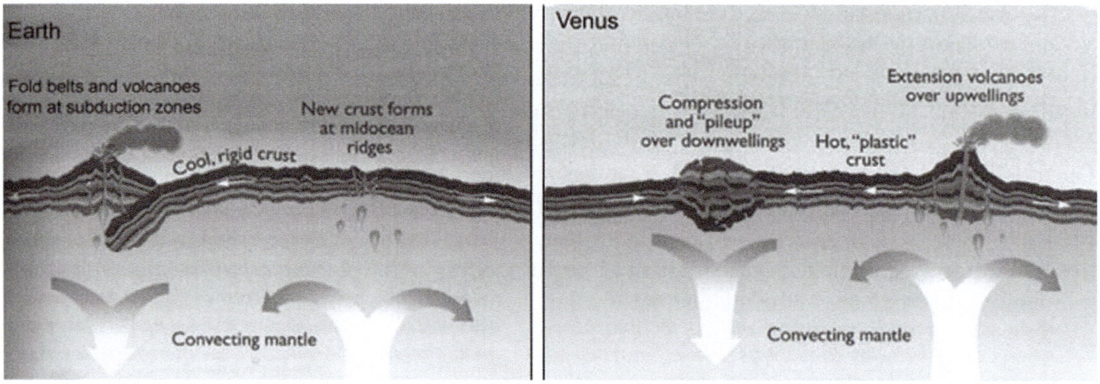

FIGURE 3.16 Comparison of tectonic model of Venus with that of Earth (from Christiansen and Spilker, 2018).

The surface of Venus appears to have been shaped by extensive volcanic activity. Venus has 167 large volcanoes that are over 100 km across. The highest point on Venus is Maxwell Montes. Magellan gathered much finer details of the surface topography of Venus, and approximately 1000 impact craters were found. Interestingly, none of the craters seen were less than 2 km in diameter. This suggests that any meteoroid small enough to create a crater having a diameter less than 2 km would have broken apart and burned up during its passage through the dense Venusian atmosphere.

The Ishtar Terra mountains formed as a result of the uplift of the plateau and formation of the Maxwell mountains, which rise 6 km above Lakshmi Planum. Low-lying areas and craters became filled with lava. The shield in the Beta Regis area was formed before the mountain ridges along the fault, which broke the surface of the shield.

The surface of Venus has shown that a wide variety of tectonic features are ubiquitous on the planet. Its surface is strongly deformed at a variety of scales. Its lithosphere has been extended and compressed, domed and depressed. Broad crustal domes and rift valleys are common tectonic features on Venus. Beta Regio is a large domical upland about 2500 km across that is crisscrossed by many faults. The gentle rise is about 4 km high, and it is crossed by a central trough. A multitude of nearly parallel linear scarps show that the depression is a fault-bounded rift valley, formed as the dome was pulled apart by extension of the lithosphere. Extensional tectonism has produced long belts of deformation, marked by abundant fractures and grabens, which persist over hundreds of kilometers. The region around Beta Regio and Atla Regio has fracture belts of diverse orientation. Because of the extensional nature and abundance in the uplands, it is inferred that they form in response to uplift of the lithosphere over mantle plumes.

3.4 MARS

3.4.1 Interior Structure

The radius of Mars is 3389 km, which is almost half the radius of Earth, and its gravitational acceleration at ground surface is 0.377g of Earth's (Watters 1995). The mean density of Mars is about 3.9335 g/cm^3 on the basis of the law of the universal gravitation. Mars has a very thin atmosphere consisting of carbon dioxides, nitrogen, argon gases, and it has a dusty, cold, desert environment. The rotation axis is tilted by 25 degrees with respect to the plane orbit around the Sun, and the year of Mars lasts about 687 Earth-days and 669.6 solar days (abbreviated as sols). Its rotation about its axis takes about 24.6 hours.

Many models for the inner structure and temperature distribution of Mars have been proposed by various researchers (e.g., Sohl and Spohn 1997; Johnston and Toksöz 1977; Toksöz and Hsui 1978). Particularly, Sohl and Spohn (1997) utilized two meteorites from Mars for estimating Mars's internal structure. All models are based on the concepts developed for Earth. Nevertheless, they are fundamentally estimations as there is no seismic network to explore Mars yet. NASA deployed a three-component seismometer through Viking program in 1976 (Anderson et al. 1977). The second seismometer was installed on Mars in 2018 as a part of the InSight instrument, which has been equipped with many sensors. The Viking seismometer recorded numerous vibrations during its operation. The InSight seismometer recorded its first marsquake in April 2019. Nevertheless, the models for the inner structure of Mars still need some validation.

The crustal thickness of Mars is about 10–50 km. The mantle is divided mainly into two layers, called the upper and lower mantles. The core–mantle interface is estimated to be at a depth of 2000 km with a 150 km transition zone. Mars is estimated to be chemically composed of a silicate crust, Fe-Mg silicates in the mantle and metallic Fe in the core (Figure 3.17). The average density is about 3.933 g/cm^3 with a gravitational acceleration of 0.377 g. Figure 3.17 illustrates the interior of Mars.

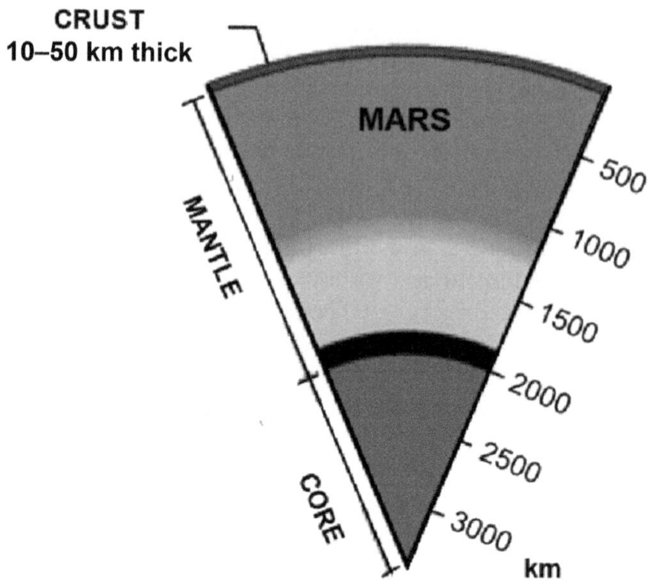

FIGURE 3.17 Interior structure of Mars (modified from NASA).

3.4.2 Geology and Tectonics

The surface geology of Mars is mapped by USGS, Scott and Carr (1978) and the MOLA data of NASA (Figure 3.18). The surface geology fundamentally involves volcanics, basalt, breccia, sedimentary deposits. The tectonics of Mars are not very active compared to that of Earth. Figure 3.19 shows a map of the tectonics of Mars prepared by USGS. Nevertheless, the very large-scale Vallis rift zone, Tharsis volcanic chain, depression zones, fracture zones, faults, folding, metamorphism, discordant sedimentation, columnar jointing are found on Mars (Figures 3.20–3.22).

3.5 MERCURY

3.5.1 Interior Structure

Similar to other terrestrial planets the core, mantle and crust of Mercury are estimated to be 1800 km, 600 km, and 10–30 km thick (Figure 3.23). Nevertheless, the internal structure of Mercury is not well-known due to limited radar and stereo coverage. Mercury's core is different from any other planetary core in the Solar System. Mercury appears to have a solid silicate crust and mantle overlying a solid, iron sulfide outer core layer, a deeper liquid core layer, and possibly a solid inner core. These results have implications for how Mercury's magnetic field is generated and for understanding how the planet evolved thermally.

Earth-based radar measurements, together with Messenger's newly measured parameters of the gravity field and the characteristics of Mercury's internal magnetic field, proved the existence of an

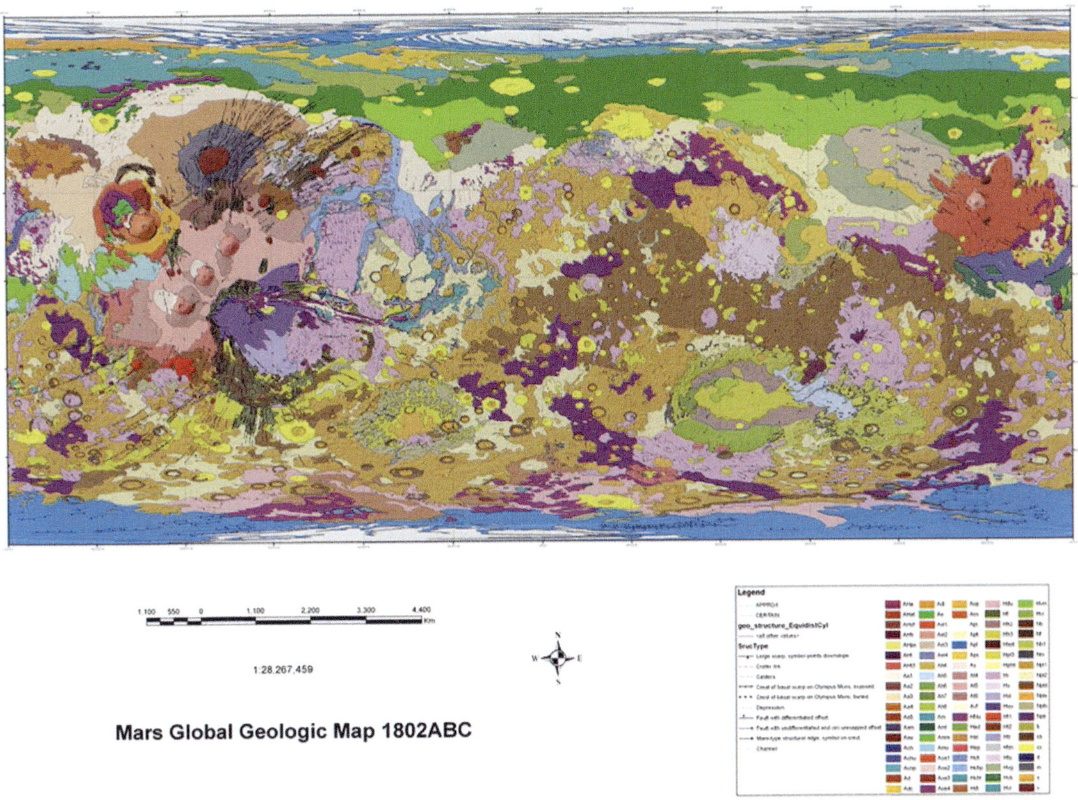

FIGURE 3.18 Geology of mars (from USGS 2020).

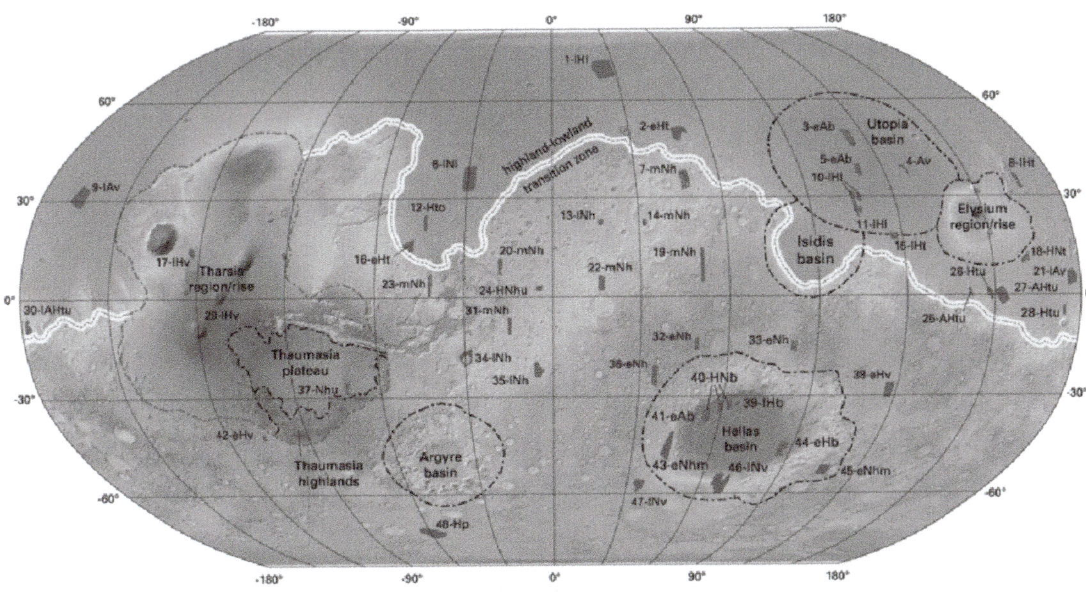

FIGURE 3.19 Topography and tectonics of Mars (from USGS 2020).

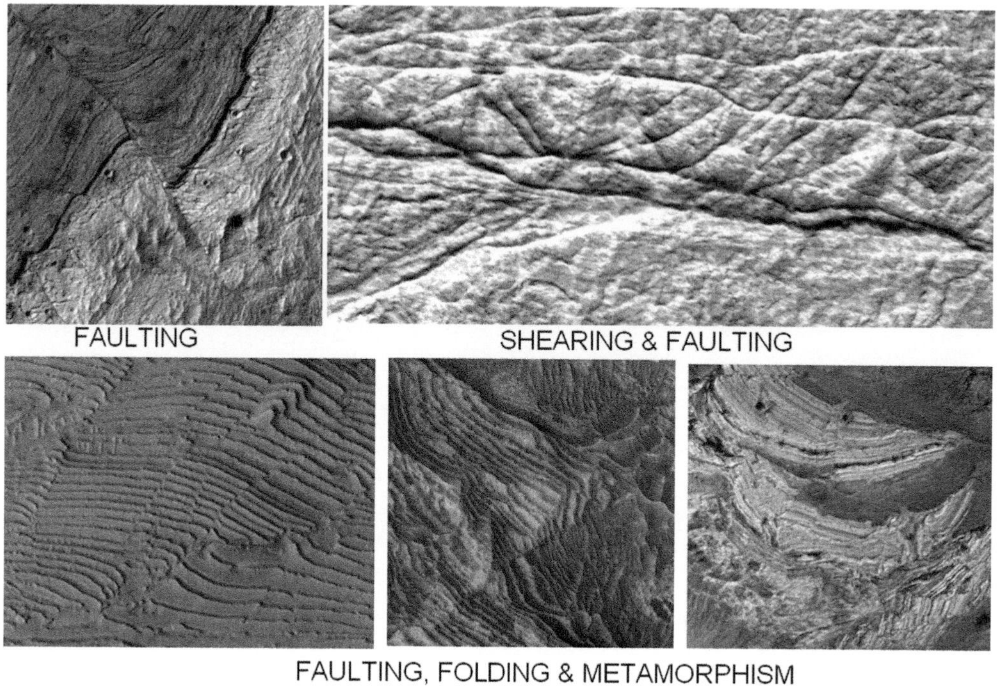

FIGURE 3.20 Large-scale faulting, shearing, folding and metamorphism structures (arranged from images by NASA).

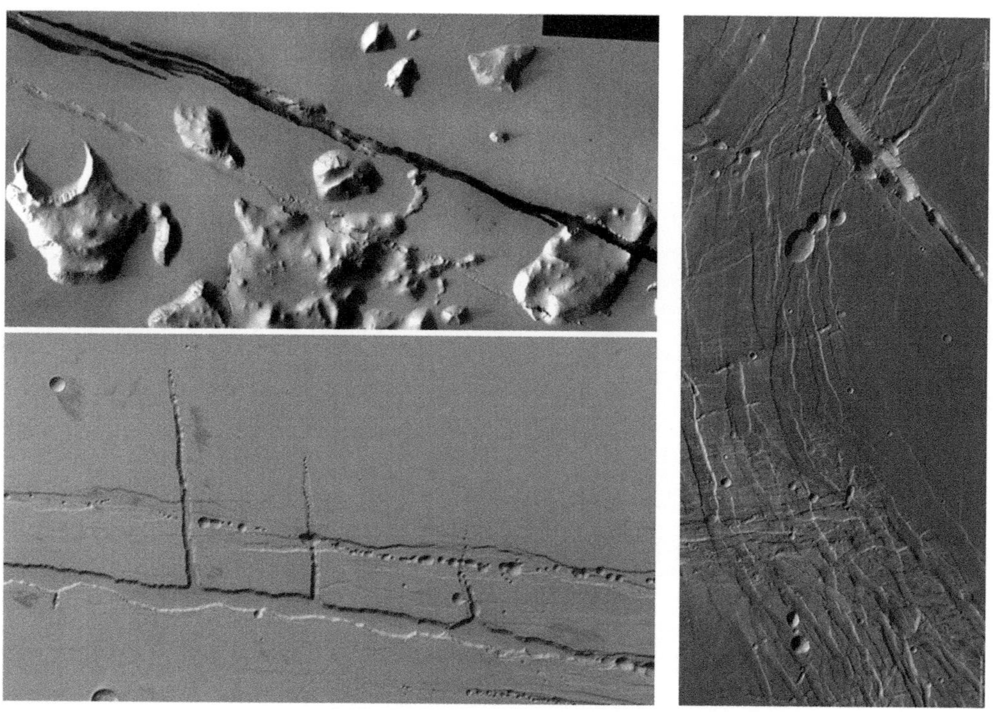

FIGURE 3.21 Large-scale fractures, rift zones (arranged from images by NASA).

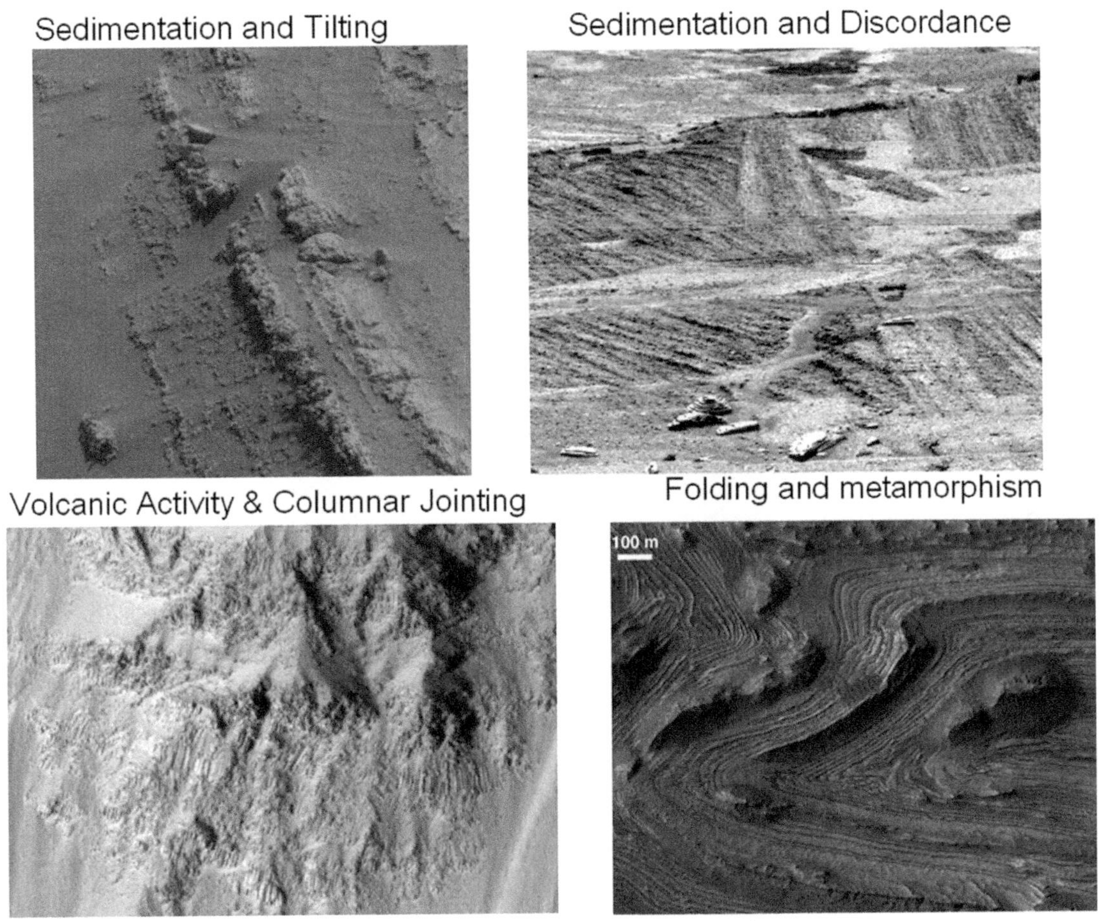

FIGURE 3.22 Sedimentation, tilting, folding, and volcanic eruptions and columnar jointing (arranged from images by NASA).

active core dynamo and indicated that the planet's core is at least partially liquid. Mercury's core occupies a large fraction of the planet, about 85% of the planetary radius, even larger than previous estimates. Because of the planet's small size, at one time many scientists thought the interior should have cooled to the point that the core would be solid.

3.5.2 Tectonics

It is a small and hot planet. Recent studies on Mercury indicated that it is contracting even in modern times. As a result of this phenomenon, shortening structures such as wrinkle ridges, lobate scarps and high-relief ridges are abundant and distributed across the surface of Mercury. It is very likely that this crustal deformation would result in quakes, mercuryquakes, and it may also imply that Mercury is a tectonically active planet. These quakes are thought to cause the formation of thrust faults and lobate scarps on the surface (Figures 3.24 and 3.25).

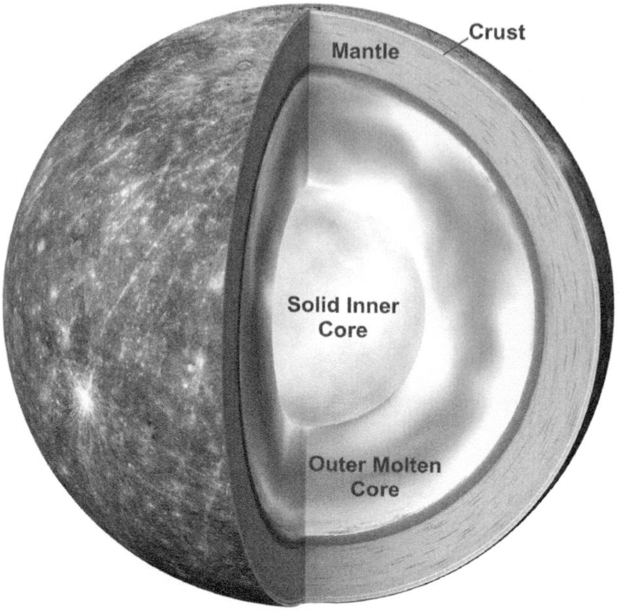

FIGURE 3.23 Estimated interior structure of Mercury (from NASA).

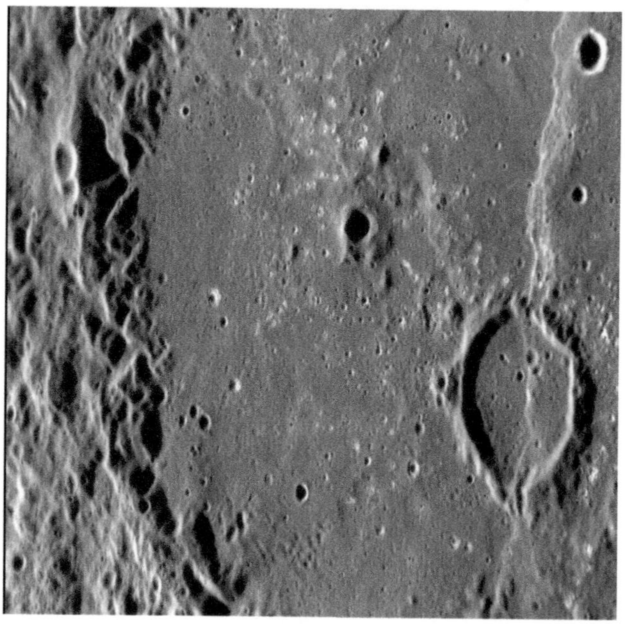

FIGURE 3.24 Oblate structure with a slight dextral movement on Mercury (from NASA),

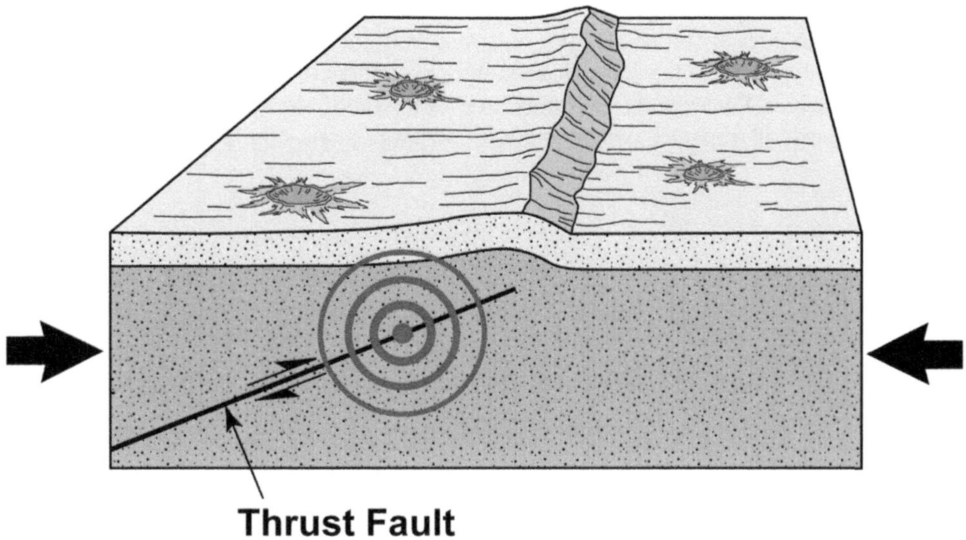

FIGURE 3.25 Estimated mechanism for oblate structures (from NASA).

Although no seismometers have been deployed on Mercury to directly measure these events, the existence of these crustal deformation structures from these surface features may confirm ongoing tectonic movements on Mercury. Topographic profiles of the lobate scarps indicate that thrust faulting extended to depths of 30–40 km, corresponding to a lower limit for the brittle–ductile transition (Watters et al. 2002).

Another estimate using wrinkle ridges in the Caloris basin implies an elastic thickness of ~100 km (Melosh and McKinnon 1988), which far exceeds the thickness inferred from the lobate scarps. As the wrinkle ridge structures were probably formed before the thrust faulting related to the lobate scarps, they cannot be explained by secular cooling (Nimmo and Watters 2004). The discrepancy between the two estimates may reflect a different curvature in these features or the local depletion of water, particularly at the surface of the fault slip surface.

Topographically, a planet can reveal fundamental information about its internal structure and its geological and thermal evolution. Ranging observations from Messenger's Mercury Laser Altimeter (MLA) have provided the first-ever precise topographic model of the planet's northern hemisphere and characterized slopes and surface roughness over a range of spatial scales. From Messenger's eccentric, near-polar orbit, the MLA illuminates surface areas as wide as 15–100 m, spaced about 400 m apart.

The spread in elevations is considerably smaller than those of Mars or the Moon. The most prominent feature is an extensive area of lowlands at high northern latitudes that hosts the volcanic northern plains. Within this lowland region is a broad topographic rise that formed after the volcanic plains were emplaced.

At midlatitudes, the interior plains of the Caloris impact basin—1550 km in diameter—have been modified so that part of the basin floor now stands higher than the rim. The elevated portion appears to be part of a quasi-linear rise that extends for approximately half the planetary circumference at midlatitudes. These features imply that large-scale changes to Mercury's topography occurred after the era of impact basin formation and large-scale emplacement of volcanic plains had ended.

Rocks and Discontinuities in Terrestrial Planetary Bodies

4

This chapter is concerned with rock and discontinuities in planetary bodies and their physico-mechanical properties. The information available on rocks and discontinuities on Earth is first presented in detail, and the information available for other planetary bodies is described in the rest of this chapter.

4.1 EARTH

The materials presented in this section are taken from a textbook by Aydan (2020a, 2020b).

4.1.1 Rocks

4.1.1.1 Igneous Rocks

Igneous rocks are created by the melting and crystallization of magma. When the magma reaches the surface, the rocks are said to be extrusive (Figure 4.1). Volcanic lava flows are examples of *extrusive igneous rocks*. If the magma cools within the Earth, it forms large bodies of crystalline rock called plutons or batholiths. These rocks are called *intrusive igneous rocks*. Igneous rocks are generally classified on the basis of three factors: (1) grain size and texture, (2) intrusive or extrusive, (3) silica content and mineral composition (Figure 4.2). However, the classifications based on factors 2 and 3 are commonly used to describe igneous rocks as follows.

4.1.1.1.1 Intrusive Igneous Rocks

These slowly cooled inside the crust. (Plutonic rock = formed in the earth.) They generally consist of large crystals of minerals.

4.1.1.1.1.1 Granite It constitutes the continental crust, and its density varies between 2.7 and 2.8 g/cm^3. It is acidic igneous rock, and it consists of quartz, mica (muscovite and biotite), potassic feldspar

DOI: 10.1201/9781003643852-4

(orthoclase, microcline), plagioclase feldspars and hornblende and tourmaline. Crystals are intermingled, and the amount of quartz is about 30%. Granitic rock is much less common on the other terrestrial planets.

4.1.1.1.1.2 Syenite Syenite is coarse-grained intrusive igneous rock of the same general composition as granite. However, the quartz is either absent or present in relatively small amounts (less than 5%). The name of this rock is related to syenites in the Syene region of Egypt.

4.1.1.1.1.3 Diorite Diorite is a gray to dark gray intermediate intrusive igneous rock composed principally of plagioclase feldspar, hornblende and/or pyroxene. The amount of quartz is greater than 10%.

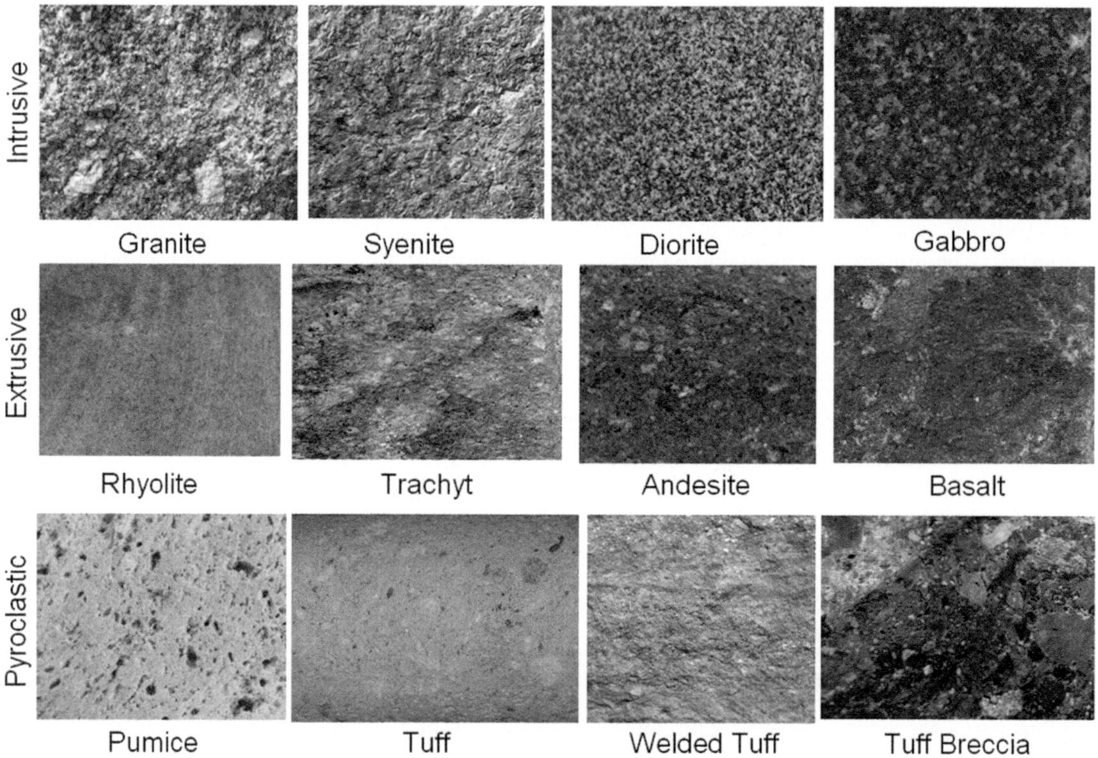

FIGURE 4.1 Views of igneous rocks.

4.1.1.1.1.4 Grano-diorite Grano-diorite is an intermediate form between granite and diorite. It usually contains abundant biotite and hornblende and orthoclase (potassium feldspar).

4.1.1.1.1.5 Gabbro It is a dark, coarse-grained, intrusive igneous rock and contains varied percentages of pyroxene, plagioclase, amphibole and olivine. The amount of quartz is less than 5%. Gabbro is generally coarse-grained, with crystals in the size range of 1 mm or greater. Gabbro is found in the ocean crust, underneath the basalt layer (0.5–2.5km) from 2.5 to 6.3 km deep.

4.1.1.1.1.6 Peridotite Peridotite is a dense, coarse-grained rock, consisting mostly of the minerals of olivine and pyroxene. Peridotite is ultramafic and ultrabasic, as the rock contains less than 45% silica. Peridotite is the dominant rock of the upper mantle of the Earth.

4.1.1.1.2 Extrusive Igneous Rocks

These rocks formed by rapid cooling of magma at the surface. As a result, crystals are either small or glassy.

4.1.1.1.2.1 Rhyolite Rhyolite has medium silica content (intermediate), and it is a fine-grained volcanic rock of granitic composition.

4.1.1.1.2.2 Trachyte Trachyte is an igneous, extrusive volcanic rock with an aphanitic to porphyritic texture. The mineral assemblage is predominately potassium feldspar with relatively minor plagioclase. It is a fine-grained volcanic rock of syenitic composition.

4.1.1.1.2.3 Andesite Andesite is considered the extrusive equivalent to diorite. The name "andesite" is derived from the Andes Mountain range.

4.1.1.1.2.4 Basalt Basalt is the extrusive equivalent to gabbro, and the two are made up of feldspars and other minerals common in planetary crusts. It constitutes the ocean crust of the Earth. It is usually fine grained due to rapid cooling of lava on the Earth's surface. Non-weathered basalt is black or gray. It has been identified as major surface rocks on the dark lunar planes and much of Mars, Venus and the asteroid Vesta.

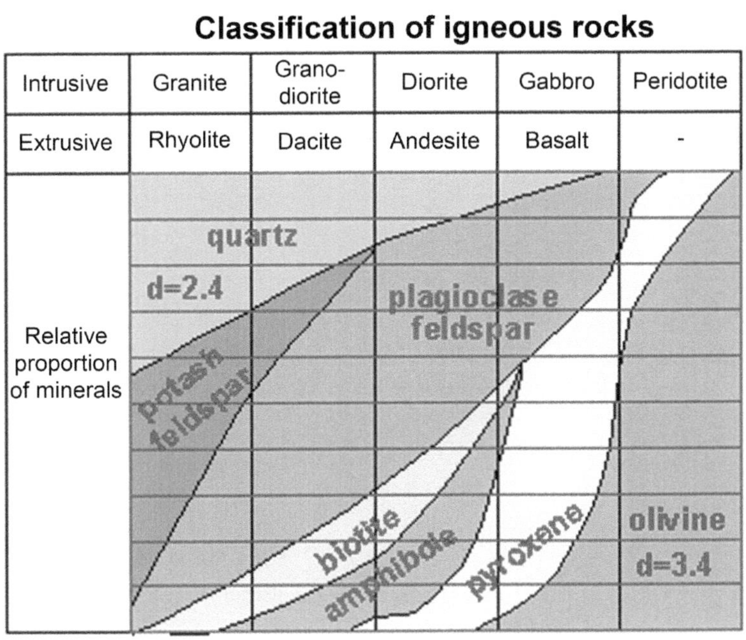

FIGURE 4.2 Chemical composition of igneous rocks (from Pinet 1992).

4.1.1.1.3 Pyroclastic Rocks

Pyroclastic rocks are formed by sedimentation and welding of debris ejected by volcanoes

4.1.1.1.3.1 Tuff Tuff is made of compacted debris from old volcanic ash showers. A tuff of recent origin is generally loose and incoherent. However, the older tuffs are cemented by pressure and the action of infiltrating water, resulting in its being strong enough yet not very hard and being extensively used for building purposes and creating cavities, as seen in the Cappadocia Region of Turkey.

4.1.1.1.3.2 Volcanic Breccia Volcanic breccia (agglomerate) is composed of angular mineral fragments embedded in a matrix, the product of explosive eruptions. Agglomerates are accumulations of large blocks of volcanic material often found around vents. Agglomerates are coarser and, less frequently, well-bedded. The blocks in agglomerates vary greatly in size.

4.1.1.1.3.3 Ignimbrites Welded tuff or ignimbrite is a product of pyroclastic flows hot enough to fuse or "weld" still-hot ash into a single uniform layer called a cooling unit. Ignimbrite is primarily composed of a matrix of volcanic ash, pumice fragments and crystals.

4.1.1.2 Sedimentary Rocks

Sedimentary rocks are formed in layers deposited by wind, water or ice. They are the direct products of the weathering process. As sedimentary layers are buried, they are cemented and lithified. Sediments are subdivided into three types (Figure 4.3):

1. Clastic sedimentary rocks
2. Chemical sedimentary rocks
3. Organic sedimentary rocks

FIGURE 4.3 Views of various sedimentary rocks.

4.1.1.2.1 Clastic Sedimentary Rocks

Clastic sedimentary rocks consist of rock and mineral grains derived from the chemical and mechanical breakdown (weathering) of preexisting rock. They contain rock fragments and, more commonly, particles of quartz and feldspar. The most common cementing materials are silica and calcium carbonate. Clastic rocks are further classified on the basis of grain size.

4.1.1.2.1.1 Conglomerates and Breccia Conglomerates are sedimentary rocks consisting of rounded fragments, while Breccia consist of angular clasts. Conglomerates and breccias are characterized by clasts larger than 2 mm.

4.1.1.2.1.2 Sandstones Sandstone is a sedimentary rock composed mainly of mineral or rock grains of sand size (0.062–2 mm). Most sandstone is composed of quartz and/or feldspar. Some sandstones are resistant to weathering and porous. When sandstone contains roughly 60% quartz sand and 25% feldspar, it is called arkose. If the percentage of quartz is greater than 95%, it is called either quartzite or quartzarenite.

4.1.1.2.1.3 Mudstone Mudstone consists of tiny particles less than 0.0062 mm in diameter. Individual grains are too small to be distinguished without a microscope. With increased pressure and time, the platy clay minerals may become aligned, with the appearance of fissility or parallel layering of consolidated mud, and is rich in organic matter.

4.1.1.2.1.4 Argillite Argillite is a sedimentary rock composed of clay particles, which have been hardened and cemented.

4.1.1.2.2 Chemical and Organic Sedimentary Rocks

These rocks are formed either from minerals that precipitate directly from aqueous (water) solutions or from the accumulation of fossilized remains of organisms, which become limestone.

4.1.1.2.2.1 Gypsum ($CaSO_4{\cdot}2H_2O$) Gypsum is a very soft chemical rock composed of calcium sulfate dihydrate.

4.1.1.2.2.2 Anhydrite ($CaSO_4$) Anhydrite, which is an evaporitic rock, is calcium sulfate. When it interacts with water, it produces an increase in volume of the rock layer. This process causes a 30% volumetric expansion, termed swelling.

4.1.1.2.2.3 Rock Salt–Halite (NaCl) Halite occurs in beds of sedimentary evaporite minerals that result from the drying up of enclosed lakes, playas, and seas. It commonly occurs with other evaporite deposit minerals such as several of the sulfates, halides and borates.

4.1.1.2.2.4 Limestone Limestone is a sedimentary rock composed largely of the calcite mineral ($CaCO_3$). Limestone may be crystalline, clastic, granular or massive, depending on the method of formation. The primary source of the calcite in limestone may be of organic origin or chemical solution. Limestone makes up about 10% of the total volume of all sedimentary rocks. Travertine is a variety of limestone formed along streams where there are waterfalls and around hot or cold springs. Limestone is partially soluble, especially in acid, and therefore forms many erosional landforms, called karsts. Tufa is a thick, rock-like calcium carbonate deposit that forms by precipitation from bodies of water with a high dissolved calcium content.

4.1.1.2.2.5 Chert Chert is a fine-grained silica-rich cryptocrystalline sedimentary rock. It varies greatly in color from white to black, but most often manifests as gray, brown, grayish brown and light green to rusty red.

4.1.1.2.2.6 Chalk Chalk is a soft, white, porous form of limestone composed of organic origin. Chalk is formed in shallow waters by the gradual accumulation of the calcite mineral remains of microorganisms such as planktonic green algae, associated with varying proportions of larger microscopic fragments of bivalves, foraminifera and ostracods.

4.1.1.3 Metamorphic Rocks

Metamorphic rocks are rocks formed by the action of pressure (P), temperature (T) and fluids within the Earth. As sediments are deeply buried, they are deformed, and new minerals recrystallize at the elevated temperatures and pressures to form metamorphic rocks. Metamorphic rocks are generated by recrystallization of either igneous or sedimentary rocks by the action of any or all of pressure, temperature and pore fluids.

The lower limit of metamorphic temperatures is 150°C. The upper limit is the melting temperature when magma forms. The type of metamorphic rock is determined by the parent rock and the P/T conditions. In general, metamorphism causes growth of new minerals, deformation and rotation of mineral grains, recrystallization of minerals and production of anisotropic rock (Table 4.1).

Contact metamorphic rocks are recrystallized and rarely show foliation (Figure 4.4). Shales baked by igneous contact form very hard fine-grained rocks called hornfels. Calcareous rocks when, subject to contact metamorphism, an alteration by hot fluids, produce rocks called skarns. Metamorphic rocks have been chemically altered by heat, pressure and deformation, while buried deep in the earth's crust. These rocks show changes in mineral composition or texture or both. This area of rock classification is highly specialized and complex.

4.1.1.3.1 Phyllite

Phyllite was originally a fine-grained sedimentary rock such as sandy mudstone or shale composed mainly of clay minerals in a semi-random orientation. Phyllite is a common metamorphic rock, and it is formed in a low-heat and high-pressure environment.

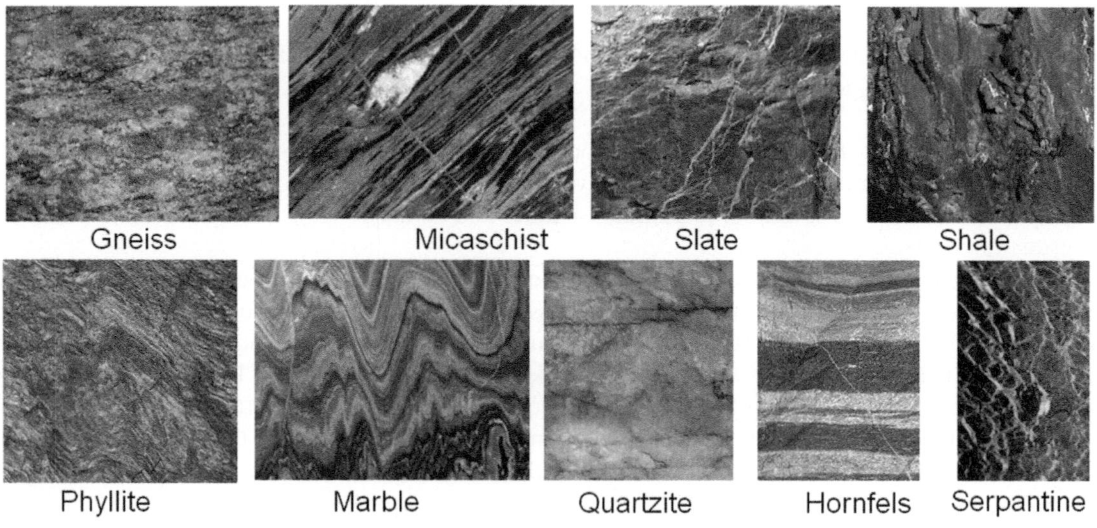

FIGURE 4.4 Views of common metamorphic rocks.

TABLE 4.1 Relationships between Parent Rock and Metamorphic Rocks

ROCK NAME	*TYPE*	*PARENT ROCK*	*CHARACTERISTICS*
Phyllite	Foliated	Silty sandstone	Splitting schistosity surfaces
Slate	Foliated	Shales and mudstone	Prominent splitting surfaces
Schist	Foliated	Fine-grained rocks	Mica minerals, often crinkled or wavy
Gneiss	Foliated	Coarse-grained rocks	Dark and light bands or layers of aligned minerals
Quartzite	Non-foliated	Sandstone	Interlocking almost fused quartz grains, little or no porosity
Marble	Non-foliated	Limestone	Interlocking almost fused calcite grains, little or no porosity
Serpentine	Non-foliated	Peridodite ultramafic rocks	

4.1.1.3.2 Shale

Shale is a fine-grained rock subjected to low-pressure metamorphism. Its original constituent is mudstone. It is characterized by thin laminae breaking with an irregular curving fracture

4.1.1.3.3 Slate

Slates are foliated rocks representing low-grade metamorphic alteration of shales (laminated clay). Slate is mainly composed of quartz and muscovite or illite, often along with other minerals.

4.1.1.3.4 Schists

Schists are foliated medium-grade metamorphic rock with parallel layers, vertical to the direction of compaction. The schists contain lamellar minerals such as micas, chlorite, talc, hornblende, graphite, and others. Quartz often occurs in drawn-out grains to such an extent that a particular form called quartz schist is produced. Schist contains more than 50% platy and elongated minerals, often finely interleaved with quartz and feldspar. Schist is characteristically foliated so that individual mineral grains split off easily into flakes or slabs.

4.1.1.3.5 Gneiss

Gneiss comes in the form of banded rocks consisting of alternating layers of quartz and feldspar, of high metamorphic grade. The original rock formations may be igneous or sedimentary rocks.

4.1.1.3.6 Quartzites

Quartzites represent metamorphosed quartzitic sandstone. Pure quartzite is usually white to gray. They often occur in various shades of pink and red due to varying amounts of iron oxide. Quartzite is very resistant to chemical weathering.

4.1.1.3.7 Marble

Marble is metamorphosed limestone, and it is composed mostly of calcite. The metamorphism causes a complete recrystallization of the original rock into an interlocking mosaic of calcite, aragonite and/or dolomite crystals.

4.1.1.3.8 Serpentinite

Serpentinite is comprised of serpentine minerals. Minerals in this group are formed by a hydration and metamorphic transformation of ultramafic rocks. Serpentinite is formed from olivine via several reactions. It can be easily weathered, resulting in swelling.

4.1.2 Discontinuities

Rocks in nature are geologically classified into three main groups; igneous rocks, sedimentary rocks and metamorphic rocks, each of which rock classes may be further subdivided into several subclasses. For example, igneous rocks are subdivided into three classes: extrusive, intrusive and semi-intrusive rocks (such as dykes, sills etc.), even though the chemical composition of the three types of rocks may be same. The order of minerals and the internal structure of rocks are a result of the chemical composition of rising magma, its velocity and environmental conditions during the cooling process, which greatly affects the discontinuity formation in such rocks. The term "discontinuity" embraces all types of interruptions of structural integrity of rock masses, which can be classified into three categories:

- *Intrinsic Discontinuities*: Bedding plane, schistosity, flow plane
- *Volumetric Strain–Induced Discontinuities*: Sheeting, desiccation, cooling, erosion, freezing–thawing
- *Plastic deformation induced discontinuities*: Faults, fracture zones, tension (T), Riedel (R-R′) shear cracks and Skempton (P) fractures.

Discontinuities in rocks are termed cracks, fractures, joints, bedding planes, schistosity or foliation planes and faults. Discontinuities are products of some certain phenomena that rocks underwent in their geological past, and they are expected to be regularly distributed within rock mass. The discontinuities can be classified into the following four groups according to the mechanical or environmental process they underwent (Erguvanlı 1973; Yüzer and Vardar 1986; Miki 1986; Ramsay and Huber 1984; Aydan 2020a, etc.) (Figure 4.5).

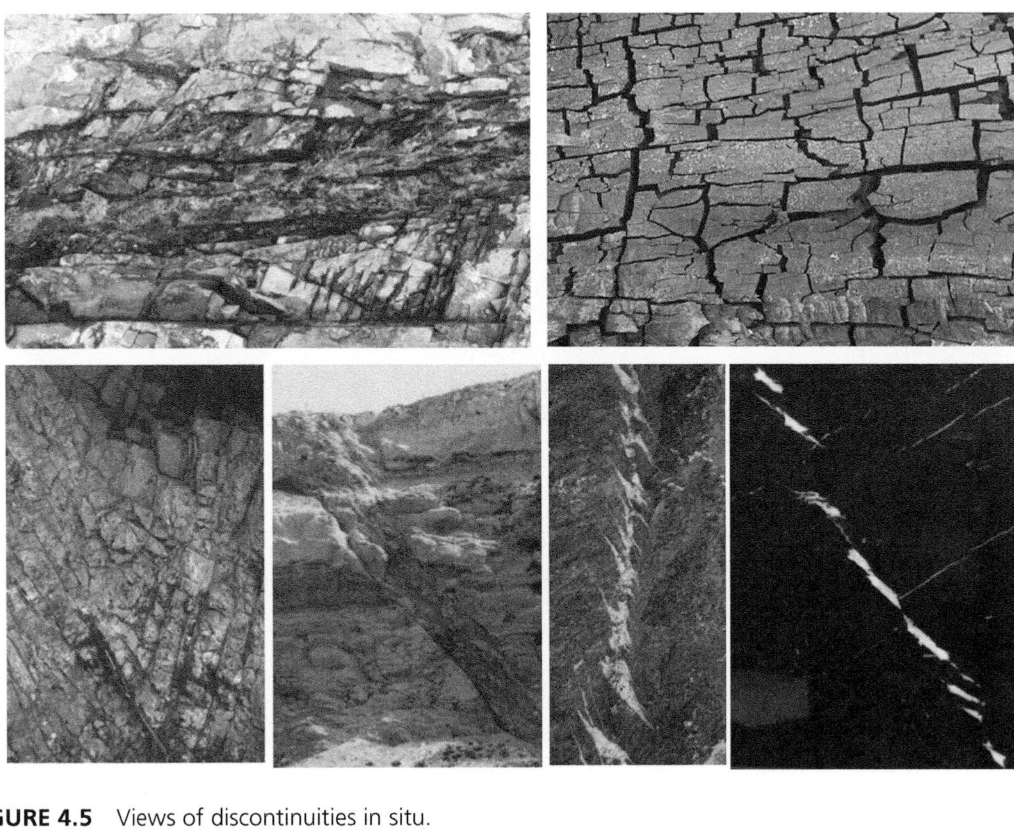

FIGURE 4.5 Views of discontinuities in situ.

1. Tension discontinuities due to
   - Cooling
   - Drying
   - Freezing
   - Bending
   - Flexural slip
   - Uplifting
   - Faulting and stress relaxation due to erosion or glacier retreat or humanmade excavation
2. Shear discontinuities due to
   - Folding
   - Faulting
3. Discontinuities due to periodic sedimentation
4. Discontinuities due to metamorphism

Discontinuities can also be healed and in-filled. Discontinuities fundamentally comprise joints, bedding, shears, contacts, veins and faults.

4.1.2.1 Joints

A joint is a discontinuity that is relatively planar and on which there has been no displacement. A series of joints in the same orientation are referred to as a joint set. Joints may be open, healed, or filled.

4.1.2.2 Bedding Joints

Bedding joints that are parallel to the bedding are referred to as bedding joints.

4.1.2.3 Foliation Joints

Foliation joints are parallel to metamorphic foliation.

4.1.2.4 Shear

Shears are structural breaks, where relative movement has occurred. The shear surfaces are characterized by the presence of slickensides, gouge, breccia, mylonite or a combination of these. Shears are in effect small faults and typically have displacements of less than 5 cm.

4.1.2.5 Fault

Fault is a shear with significant continuity and evidence of large displacement. A fault can range from centimeters in width to a zone that is tens of meters thick. The fault may contain breccia, gouge and crushed rock. Fault zones are typically conduits for high groundwater flow.

4.1.2.6 Contact

Contact is a geologic contact between two distinct lithologic units.

4.1.2.7 Vein

A vein is an infilling of a discontinuity caused by circulation of mineralized fluid and deposition of minerals. Veins can cause healing of the original discontinuity.

Discontinuities, although they may be viewed as planes in large scale, have undulated surfaces varying in irregularities. As a result, they may be regarded as bands with a certain thickness in association with the amplitude of undulations. The discontinuities may be filled up by some infilling materials such as calcite, quartzite or weathering products of host rock or transported materials, or they may exist from the beginning as thin films of clay deposits in sedimentary rocks along bedding planes.

4.2 MOON

4.2.1 Lunar Rocks

From Apollo missions, 382 kg rock samples were brought to Earth from the Moon. Figure 4.6 shows views of several examples of Moon rocks such as olivine, basalt and anorthositic plagioclase. Other rock samples are tuff breccias associated probably with meteorite impacts. Figure 4.7 shows larger rock blocks.

The lunar highlands have gabbro. Although rock samples sampled and brought to Earth by the astronauts are quite limited so far, there should be some clastic sedimentary rocks as well as metamorphic rocks associated with crater formation resulting from meteorite collision or other causes. The new explorations on the Moon may extend our knowledge on Lunar rocks.

FIGURE 4.6 Views of some Moon rocks (original images from NASA).

4.2.2 Discontinuities

One can easily notice from images taken by Apollo expeditions that bedding planes, fracture planes, cooling joints and faults do exist on the Moon, and they are very similar to those of Earth. Figures 4.8–4.10 show several examples of bedded rock mass, thrust fault and columnar jointing on the Moon. In other words, the rock mass concept and related knowledge in the field of rock mechanics and rock engineering must be also used to describe rocks of terrestrial planetary bodies rather than intact rock only.

FIGURE 4.7 Large-scale in situ rock boulders of the Moon with discontinuities.

FIGURE 4.8 Bedded rock mass on the Moon (image from NASA).

FIGURE 4.9 Thrust faulting on the Moon (image from NASA).

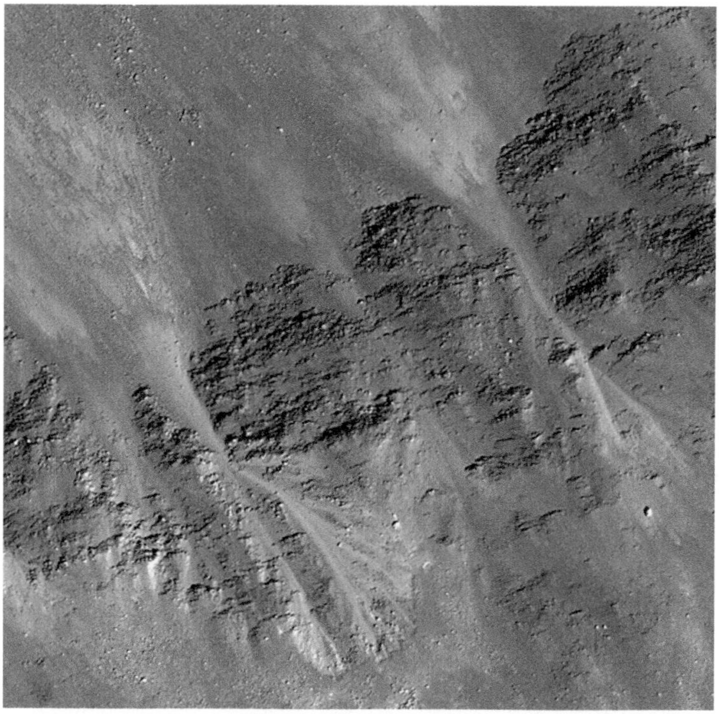

FIGURE 4.10 Columnar jointing on the Moon (Dawes Crater, from LRO, NASA).

4.3 VENUS

4.3.1 Venus Rocks

The Soviets have successfully landed several spacecraft in the plains regions (Figure 4.11). Seven of the landers conducted chemical analyses of rocks, which indicate a composition similar to that of terrestrial basaltic volcanic rocks. Venera spacecraft, which landed on Venus, took images of rocks of Venus (Figure 4.12). The thin, plate-like slabs of rock could be due to molten lava that cooled and cracked. The composition and texture of these rocks is thought to be similar to terrestrial basaltic lava.

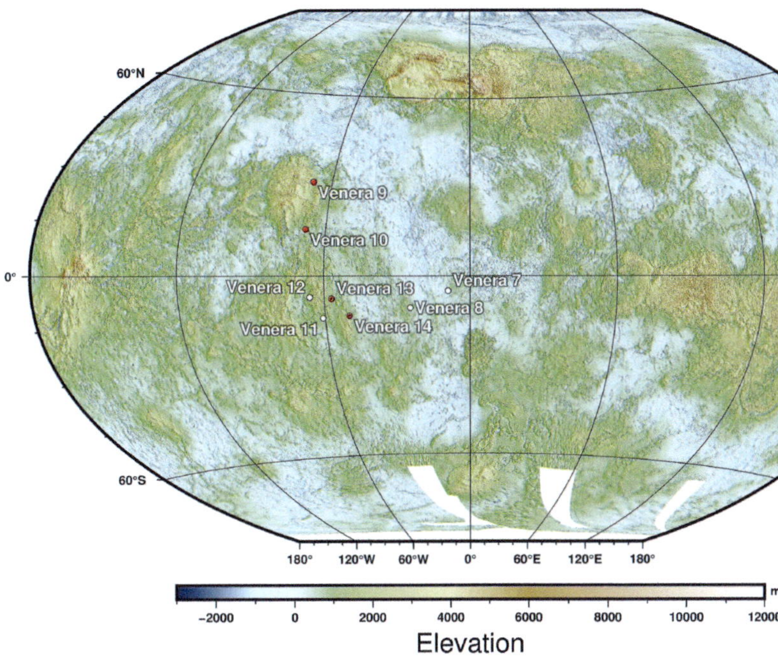

FIGURE 4.11 Landing locations of Vega and Venera spacecraft on Venus (Wikipedia-zamonin).

Weathering processes of rocks on Venus are due to the extremely high temperature and pressure (about 9 MPa) at the ground surface and the composition of the atmosphere. Each of these factors exerts some control on chemical reactions between gas and rock. Minerals that crystallized at high temperatures in lava flows and that are exposed to CO_2 and SO_2 gas in the Venusian atmosphere are inherently unstable and decompose to form new minerals. The weathered zone probably consists of a mixture of incompletely reacted minerals and newly formed weathering products. For example, it is predicted that, as weathering decomposes basaltic lavas, iron oxides and sulfur-rich minerals form as iron silicates are destroyed. However, carbonate minerals, which could remove significant amounts of carbon dioxide from the atmosphere, are not stable on the hot, dry surface of Venus.

Weathering varies with altitude on Venus. At high altitudes and low temperatures, sulfur in the atmosphere reacts with lava flows to form the mineral pyrite (FeS_2). Pyrite is highly reflective of radar waves and may explain why the mountain peaks and high plateaus of Venus are radar bright. At lower altitudes and higher temperatures, magnetite (Fe_3O_4) and anhydrite ($CaSO_4$) may be the stable minerals produced by weathering. Magnetite is not as reflective as pyrite.

4.3.2 Discontinuities

Compared to Earth, Venus is very active in terms of geological and environmental events. Furthermore, the thick clouds inhibit direct observation on Venus except some very limited observations by the

FIGURE 4.12 Images of rocks and soil around USSR Venera Spacecraft landing sites on Venus (Russian Academy of Sciences).

Venera spacecraft of Soviets. As noted from Figure 4.12, distinct layering and fracturing on the rock mass at the ground surface and in mountains do exist.

The Magellan spacecraft of NASA in 1990 explored the surface of Venus with a resolution of 100 m using the synthetic aperture radar technique. Therefore, the imaging of small-scale discontinuities is not possible Nevertheless, it was successful enough to map large-scale discontinuities such as faults, grabens, craters, highly deformed and fractured zones, large-scale slope failures as well as volcanoes and magma flows (Figures 4.13 and 4.14). It was even possible to recognize a new faulting event due to a Venusquake from flyovers of the same area between 1990 and 1994.

4.4 MARS

4.4.1 Rocks

It is very likely that the rocks of Mars would be also close to those of Earth described in subsection 4.1.1. Therefore, the classification of rocks should be igneous, metamorphic and sedimentary on the basis of the images from many Mars explorers of NASA. The images and drilling operations performed by Mars explorers (Opportunity, Spirit, Curiosity and Perseverance) on some selected rocks and rock blocks indicated that rocks are quite similar to those of the earth. Although it still needs some

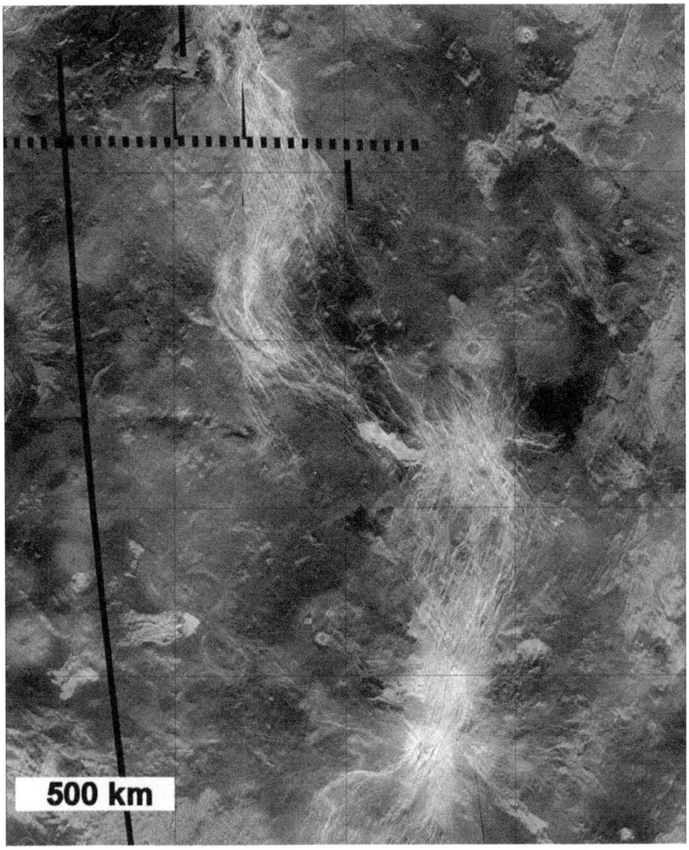

FIGURE 4.13 Long fractured zone in Devana Chasma of Venus (from NASA).

FIGURE 4.14 Image of an active volcano (Maat Mons–NASA).

clarifications whether rocks seen in Earth fully exist on Mars, the sedimentary rocks (such as conglomerate, sandstone, siltstone, mudstone, sulfate deposits), metamorphic rocks such as phyllite, shale and serpentine and extrusive rocks (basalt, andesite-like [silicate bearing basalt], breccia) of Mars are clearly similar to those of Earth (Figures 4.15–4.17). Perseverance explorer recently came across two rock samples resembling gneiss and granite as seen in Figure 4.18.

(a) Basalt (b) Andesite (c) Grano-diorite

FIGURE 4.15 Views of intrusive and extrusive igneous rocks (arranged from images from NASA).

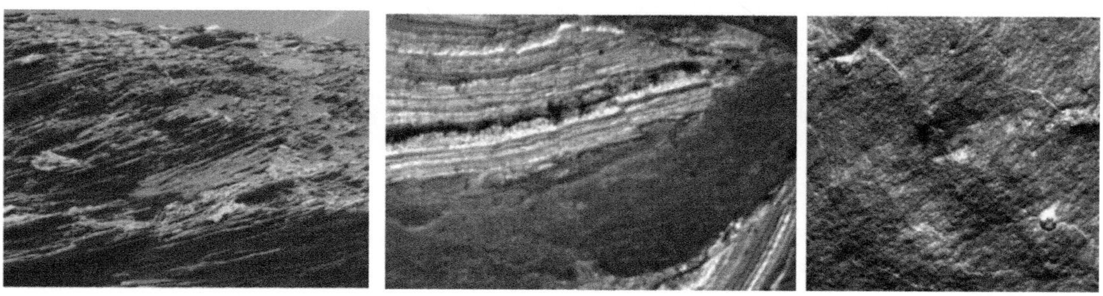

(a) Shale (b) Folded schist (c) Green-schist with garnet

FIGURE 4.16 Views of metamorphic rocks (arranged from images from NASA).

(a) Conglomerate (b) Anhydrite (c) Mudstone

FIGURE 4.17 Views of sedimentary rocks (arranged from images from NASA).

(a) Gneiss-like (b) Granite-like

FIGURE 4.18 Rock samples resembling gneiss and granite found by Perseverance rover (arranged from images from NASA).

4.4.2 Discontinuities

Flow planes, bedding planes and schistosity planes, which are intrinsic to each rock class, are distinctly observed. In addition, some discordant sedimentation of sandstone layers was also distinctly observed.

As explained previously, there are various causes for the formation of discontinuities in rocks. The causes can be grouped into (1) tension discontinuities due to cooling, drying, freezing, bending, flexural slip, uplifting, faulting, and stress relaxation associated with erosion or retreating glacier; (2) shear discontinuities due to folding and faulting; (3) discontinuities due to the intrinsic properties of rocks (sedimentation, metamorphism, magma flow). Without exception, discontinuities in rocks/rock masses of Mars are fundamentally similar to those observed in Earth, as seen in Figure 4.19.

The surface morphology and filling situations of rock discontinuities are quite similar to those seen in Earth (Figure 4.20). Although the striation of shear discontinuities could not be confirmed from the images taken by the Mars Exploration Rovers, they should be quite similar to those observed on Earth. The chemical analyses of the filling material so far showed that they generally consist of calcium sulfate (anhydrite) or calcium carbonate (calcite). Sometimes they include boron minerals, which are probably associated with dissolved substances when the planet had a wet environment in the past.

Filling of discontinuities is expected to improve the mechanical properties of discontinuities.

4.4.3 Weathering of Rocks in Mars

It is well-known that the environment of Mars was quite different from the present one in view of the surface morphology of Mars. There were lakes, rivers and oceans, and climatic conditions were entirely different from the present. Images from Mars explorers clearly show that the weathering of the rocks of mars is quite similar to those on Earth. As seen in the images of Figure 4.21, differential weathering, solution and oxidation type weathering can be easily noticed.

Compared to the daily temperature difference on the Moon, the daily temperature difference in Mars is about 80°C. If the rock were not saturated, such a temperature difference would not be of great significance, except for soft rocks. Nevertheless, the crumbling of rocks may take place due to cyclic temperature variations.

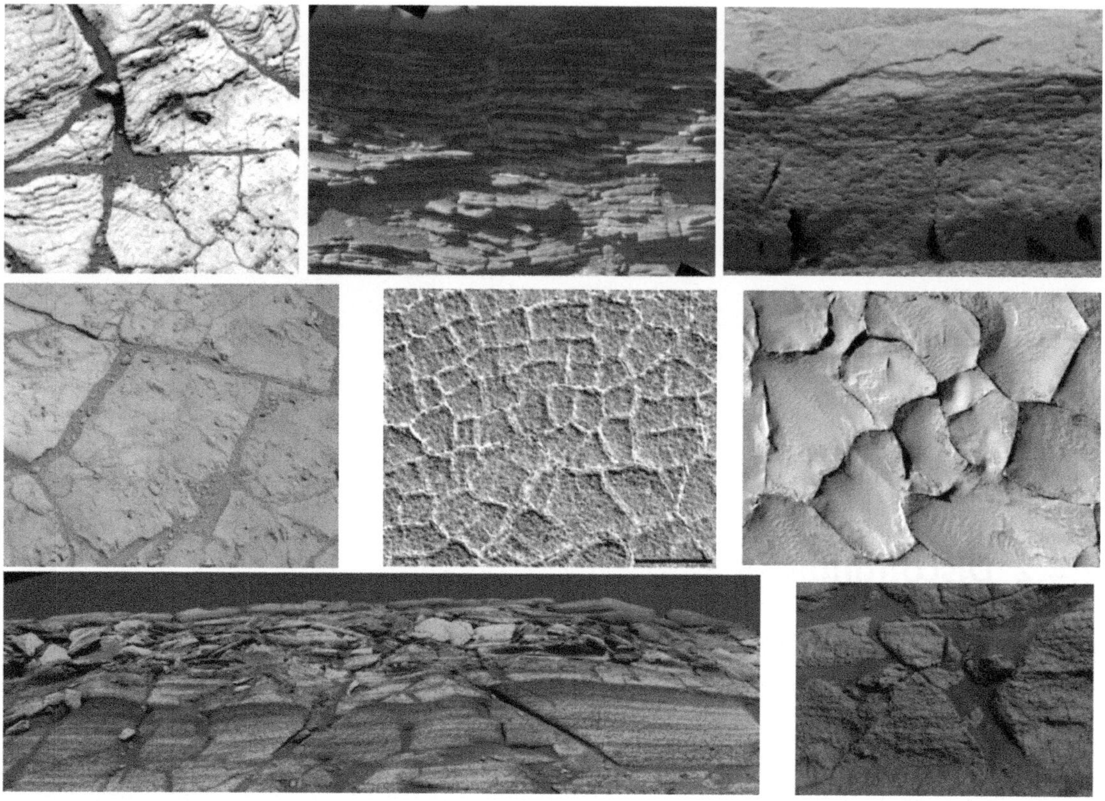

FIGURE 4.19 Discontinuities in rocks/rock masses of Mars.

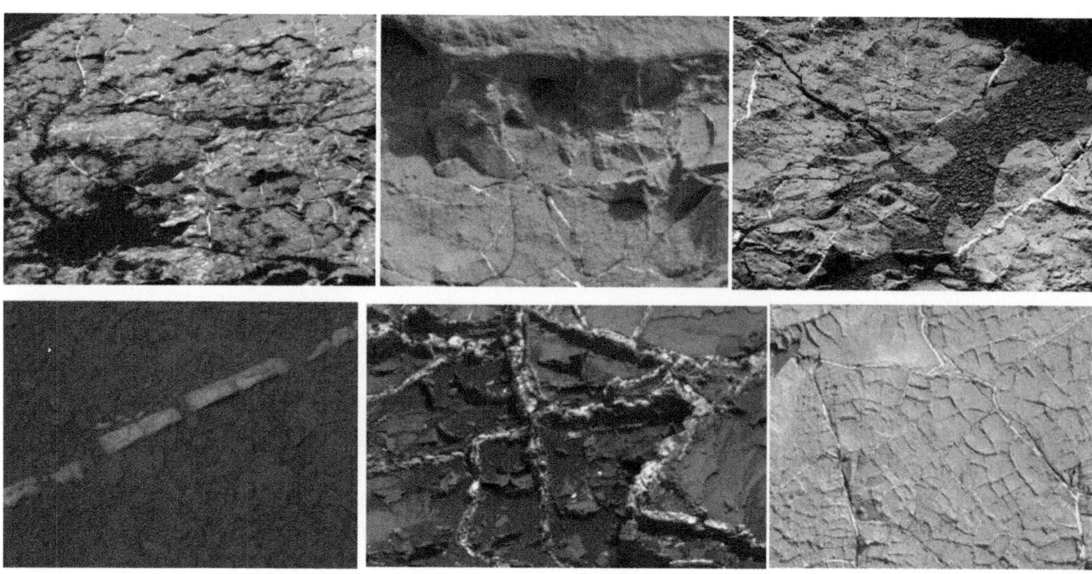

FIGURE 4.20 Images of discontinuity filling in rocks of Mars.

FIGURE 4.21 Weathering of rocks of Mars.

4.5 MERCURY

4.5.1 Rocks

Presently no explorers are landed on Mercury. Therefore, it is very difficult to make statements firmly about the rocks and their types. The surface rocks are presumed to be composed of materials similar to terrestrial basalts and komatiites. The composition likely to include minerals of plagioclase, pyroxene, olivine and possibly sulfides. In addition, pyroclastic rocks are found particularly nearby impact craters and volcanos. The existence of impact craters and volcanos may also imply the existence of metamorphic rocks.

FIGURE 4.22 Oblique view of Mercury.

4.5.2 Discontinuities

The surface of Mercury is heavily cratered, with an extensive oblate system caused by thrust faulting (Figure 4.22). This oblate structure is thought to be due to the shrinkage of Mercury during the cooling process. It has a bright ray system resulting from impact ejecta. The fault inclination is likely to be changing between 20 and 45 degrees. If the faults are frictional, they may dip about 30 degrees.

Fundamental Laws, Constitutive Models and Testing

5

The fundamental laws are the laws of continuum mechanics (e.g., Eringen 1980; Mase 1970), and they can be utilized not only for Earth but also other terrestrial bodies. Although the derivation of the fundamental laws is not presented here, the readers are recommended to see some textbooks (e.g., Eringen 1980; Mase 1970; Aydan 2017a, 2017b, 2020a, 2021c)

5.1 FUNDAMENTAL GOVERNING LAWS

5.1.1 Mass Conservation Law

The governing equation for mass conservation may be written as

$$\frac{d\rho}{dt} + \nabla \cdot \mathbf{q} = \pm Q \tag{5.1}$$

Where $\rho, \mathbf{q}$ and Q are density, flux vector and source or sink, respectively. ∇ is the directional derivative operator.

5.1.2 Momentum Conservation Law

Momentum conservation law for rock mass can be derived as is done in continuum mechanics, and the final form is:

$$\rho \frac{\partial \mathbf{v}}{\partial t} = \nabla \cdot \boldsymbol{\sigma} + \mathbf{b} \tag{5.2}$$

where $\rho, \mathbf{v}, \boldsymbol{\sigma}$ and $\mathbf{b}$ are density, velocity, stress tensor and body force, respectively.

DOI: 10.1201/9781003643852-5

5.1.3 Angular Momentum Conservation Law

The final form of the angular momentum equation implies that stress tensor is a symmetric tensor, that is:

$$\sigma_{ij} = \sigma_{ji} \tag{5.3}$$

5.1.4 Energy Conservation Law

Energy conservation law of rock mass may be written in the following form:

$$\rho c \frac{\partial T}{\partial t} = -\nabla \cdot \mathbf{q}_h + \boldsymbol{\sigma} \cdot \dot{\varepsilon} + Q_h \tag{5.4}$$

where $c, T, \mathbf{q}_h, \dot{\varepsilon}$ and Q_h are specific heat, temperature, strain rate tensor and heat source or sink, respectively.

5.2 DEFORMATION AND STRAIN, STRAIN RATE RELATIONS

The relation between deformation and strain is derived using Lagrangian and Eulerian descriptions, and they can be related to a small strain tensor. Figure 5.1 illustrates the main concept in a two-dimensional situation for deriving the relations.

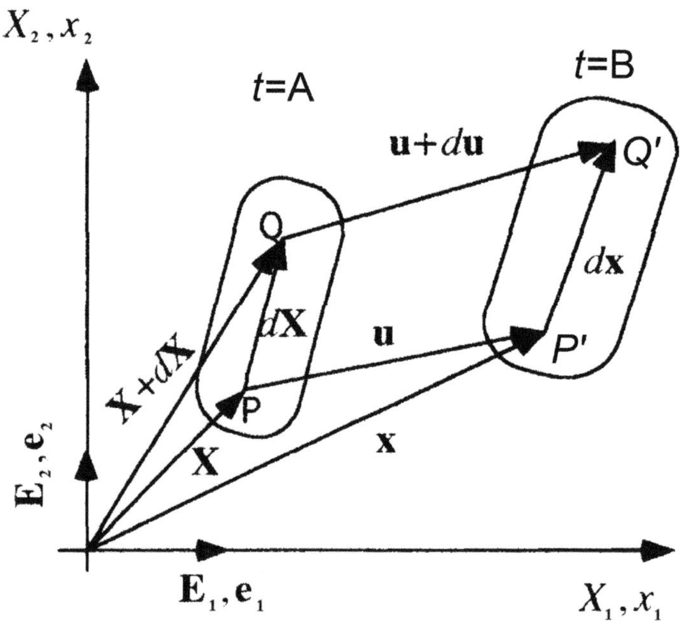

FIGURE 5.1 Coordinate system and notations.

For Lagrangian description, one can derive the following relation:

$$\mathbf{L} = \frac{1}{2}\left[(\nabla_X \mathbf{u}) + (\nabla_X \mathbf{u})^T + (\nabla_X \mathbf{u})^T \cdot (\nabla_X \mathbf{u})\right] \tag{5.5}$$

Where ∇_X and $\mathbf{u}$ are directional derivation operator and deformation vector in Lagrangian description. The Lagrangian strain tensor is also known as the Green strain tensor. It is interpreted as a finite strain tensor. In index notation, Eq. (5.5) is expressed as follows:

$$L_{jk} = \frac{1}{2}\left[\frac{\partial u_j}{\partial X_k} + \frac{\partial u_k}{\partial X_j} + \frac{\partial u_i}{\partial X_j}\frac{\partial u_i}{\partial X_k}\right] = \frac{1}{2}\left[u_{j,k} + u_{k,j} + u_{i,j}u_{i,k}\right] \tag{5.6}$$

As for Eulerian description, the following relations can be derived.

$$\mathbf{E} = \frac{1}{2}\left[(\nabla_x \mathbf{u}) + (\nabla_x \mathbf{u})^T - (\nabla_x \mathbf{u})^T \cdot (\nabla_x \mathbf{u})\right] \tag{5.7}$$

The Eulerian strain tensor is also known as the Almani strain tensor, and it is the finite strain tensor used for large deformations of materials. In index notation, it is rewritten as:

$$E_{jk} = \frac{1}{2}\left[\frac{\partial u_j}{\partial x_k} + \frac{\partial u_k}{\partial x_j} - \frac{\partial u_i}{\partial x_j}\frac{\partial u_i}{\partial x_k}\right] = \frac{1}{2}\left[u_{j,k} + u_{k,j} - u_{i,j}u_{i,k}\right] \tag{5.8}$$

As noted from strain definitions given by Eqs. (5.6) and (5.8), the term corresponding to the power of strain components is noted. Therefore, the finite strain tensors are geometrically nonlinear, and their use in practice becomes troublesome. When the strain is small, say less than 10%, the nonlinear components may be omitted. Furthermore, if coordinate systems are assumed to be the same, say, $\mathbf{x} = \mathbf{X}$, strain tensors given by Eqs. (5.6) and (5.8), reduced to the following form:

$$E_{jk} = L_{jk} = \frac{1}{2}\left[\frac{\partial u_j}{\partial x_k} + \frac{\partial u_k}{\partial x_j}\right] = \frac{1}{2}\left[u_{j,k} + u_{k,j}\right] \tag{5.9}$$

The strain tensor given by Eq. (5.9) is the most commonly used form in practice.

5.3 CONSTITUTIVE LAWS

5.3.1 Linear Laws

5.3.1.1 Fourier's Law

Fourier's law states that heat flux q_i is linearly proportional to the gradient of temperature T, that is:

$$q_i = -K_{ij}\frac{\partial T}{\partial x_j} \tag{5.10}$$

where K_{ij} is called the thermal heat conductivity tensor.

5.3.1.2 Fick's Law

Fick's law is essentially the same as that of Fourier's, and it is used in diffusion problems. Fick's law states that the mass flux q_i is linearly proportional to the gradient of mass concentration C, that is:

$$q_i = -D_{ij} \frac{\partial C}{\partial x_j} \tag{5.11}$$

where D_{ij} is called the diffusion coefficient tensor.

5.3.1.3 Darcy's Law

Darcy's law is essentially the same as that of Fourier's, and it is used in groundwater seepage problems. Darcy's law states that seepage flux q_i is linearly proportional to the gradient of ground water pressure p, that is:

$$q_i = -K_{ij} \frac{\partial p}{\partial x_j} \tag{5.12}$$

where K_{ij} is called the hydraulic conductivity tensor.

5.3.1.4 Hooke's Law

When rock or rock mass behaves linearly without any rate dependency, the simplest constitutive law is Hooke's law. This law is written in the following form:

$$\sigma_{ij} = D_{ijkl}\varepsilon_{kl} \tag{5.13}$$

where $\sigma_{ij}, \varepsilon_{kl}$ and D_{ijkl} are stress, strain and elasticity tensors, respectively.

If material is homogeneous and isotropic, Eq. (5.13) may be written as

$$\sigma_{ij} = 2\mu\varepsilon_{ij} + \lambda\delta_{ij}\varepsilon_{kk} \tag{5.14}$$

where δ_{ij} is Kronecker delta tensor. λ and μ are Lame coefficients, which are given in terms of elasticity (Young's) modulus (E) and Poisson's ratio (υ) as:

$$\lambda = \frac{E\upsilon}{(1+\upsilon)(1-2\upsilon)};\ \mu = \frac{E}{2(1+\upsilon)} \tag{5.15}$$

5.3.1.5 Newton's Law

Newton's law is used in the fluid mechanics. Newton's law states that stress σ_{ij} is linearly proportional to strain rate $\dot{\varepsilon}_{kl}$, that is:

$$\sigma_{ij} = C_{ijkl}\dot{\varepsilon}_{kl} \tag{5.16}$$

where C_{ijkl} is the viscosity tensor.

5.3.1.6 Kelvin-Voigt's Law

Kelvin-Voigt's law is used in the field of visco-elasticity. Voigt's law states that stress σ_{ij} is linearly proportional to strain ε_{ij} and strain rate $\dot{\varepsilon}_{ij}$, that is:

$$\sigma_{ij} = D_{ijkl}\varepsilon_{kl} + C_{ijkl}\dot{\varepsilon}_{kl} \tag{5.17}$$

where $\dot{\varepsilon}_{kl}$ and C_{ijkl} are strain rate and viscosity tensors, respectively.

If material is homogeneous and isotropic, Eq. (5.17) may be written in analogy to Eq. (5.14) as

$$\sigma_{ij} = 2\mu\varepsilon_{ij} + \lambda\delta_{ij}\varepsilon_{kk} + 2\mu^*\dot{\varepsilon}_{ij} + \lambda^*\delta_{ij}\dot{\varepsilon}_{kk} \tag{5.18}$$

Coefficients λ^* and μ^* may be called viscous Lame coefficients.

5.3.1.7 Navier–Stokes Law

Navier–Stokes constitutive law can be visualized as a simple case of Kelvin–Voigt's law. If the material behavior is associated with volumetric strain (ε_v) and pressure (p) under the steady-state condition without any shear resistance like fluids, Eq. (5.18) can be rewritten as:

$$\sigma_{ij} = -p\delta_{ij} + 2\mu^*\dot{\varepsilon}_{ij} + \lambda^*\delta_{ij}\dot{\varepsilon}_{kk} \tag{5.19}$$

If coefficient $\lambda^* = 0$, the preceding relation reduces to the following form:

$$\sigma_{ij} = -p\delta_{ij} + 2\mu^*\dot{\varepsilon}_{ij} \text{ or } \sigma_{ij} = -p\delta_{ij} + \mu^*\dot{\gamma}_{ij} \tag{5.20}$$

This constitutive law corresponds to the constitutive law commonly used in fluid mechanics known as Navier–Stokes law.

There are different visco-elasticity models in the literature. As it is difficult to cover all the models here, the reader is advised to consult the available literature on the topic of visco-elasticity.

5.3.2 Elasto-Plasticity and Elasto-visco-plasticity for Solids

Every material in nature starts to yield after a certain stress or strain state, and rock or rock mass is a no exception. The terms used to describe the material behavior such as elasticity and visco-elasticity are replaced by the terms of elasto-plasticity or elasto-visco-plasticity as soon as material behavior deviates from linearity. The relation between total stress and strain or strain rate can no longer be used, and every relation must be written in incremental form. For example, if the conventional plasticity models are used, the elasto-plastic constitutive law between incremental stress and strain tensors would take a different form.

5.3.2.1 Elasto-Plasticity

When the material behavior becomes nonlinear, the total strain and total stress relations cannot be used anymore. Therefore, all relations are given in incremental form as:

$$d\sigma = \left(\mathbf{D}^e - \frac{d\mathbf{D}^e \dfrac{\partial G}{\partial \sigma} \otimes \dfrac{\partial F}{\partial \sigma} \mathbf{D}^e}{h + \dfrac{\partial F}{\partial \sigma} \cdot \left(\mathbf{D}^e \dfrac{\partial G}{\partial \sigma} \right)} \right) d\varepsilon \tag{5.21}$$

The hardening modulus h is generally determined as a function of a hardening parameter κ by employing either a work-hardening model or strain-hardening model. G and F are plastic potential and yield functions, respectively. The hardening parameter κ is defined for both cases as follows:

$$\kappa = W^p = \int \boldsymbol{\sigma} \cdot d\boldsymbol{\varepsilon}^p \quad \text{work-hardening} \tag{5.22}$$

$$\kappa = \int \| d\boldsymbol{\varepsilon}^p \| \quad \text{strain hardening} \tag{5.23}$$

where W^p is plastic work.

The materials (i.e., steel, glass fibers) exhibit a non-dilatant plastic behavior and isotropically harden. Therefore, a work-hardening model is generally used together with the effective stress–strain concept, which are defined as:

$$\sigma_e = \sqrt{\frac{3}{2}(\mathbf{s} \cdot \mathbf{s})} \quad d\varepsilon_e^p = \sqrt{\frac{2}{3}(d\boldsymbol{\varepsilon}^p \cdot d\boldsymbol{\varepsilon}^p)} \tag{5.24}$$

where $\mathbf{s}$ is deviatoric stress tensor.

As the volumetric plastic strain increment $d\bar{\varepsilon}_v^p = 0$, together with the coaxiality of the stress and plastic strain, the hardening parameter κ of work-hardening type can be rewritten in the following form:

$$d\kappa = dW^p = \boldsymbol{\sigma} \cdot d\boldsymbol{\varepsilon}^p = \mathbf{s} \cdot d\mathbf{e}^p = \sigma_e d\varepsilon_e^p \tag{5.25}$$

where $d\mathbf{e}^p$ is the deviatoric strain increment. The hardening modulus h for this case takes the following form with the use of Euler's theorem[1] by taking a homogeneous plastic potential G of order m:

$$h = -\frac{\partial F}{\partial \kappa}\frac{\partial \kappa}{\partial \boldsymbol{\varepsilon}^p} \cdot \frac{\partial G}{\partial \boldsymbol{\sigma}} = \frac{\partial K}{\partial W^p}\boldsymbol{\sigma} \cdot \frac{\partial G}{\partial \boldsymbol{\sigma}} = m\frac{\partial K}{\partial W^p}G \tag{5.26}$$

If $F = G$ and $f(\boldsymbol{\sigma}) = \sigma_e$, then the hardening modulus h becomes:

$$h = m\frac{\partial K}{\partial W^p} f(\sigma) = m\frac{\partial \sigma_e}{\partial \varepsilon_e} \tag{5.27}$$

The hardening modulus can then be easily obtained from a gradient of the plot of a uniaxial test in σ_1 and ε_1^p space, since the effective stress and strain in the uniaxial state become:

$$\sigma_e = \sigma_1 \quad \varepsilon_e^p = \varepsilon_1^p \tag{5.28}$$

[1] $\mathbf{x} \cdot \partial f / \partial \mathbf{x} = mf$

5.3.2.2 *Elasto-visco-plasticity Models*

These approaches assume that the materials are elastic before yielding and behave in a visco-plastic manner following yielding. In visco-plastic evaluations, $\mathbf{e}_p$ is replaced by $\mathbf{e}_{vp}$.

5.3.2.2.1 Power-type Models

When Norton-type constitutive law is used for creep response, the visco-plastic strain rate ($\mathbf{e}_{vp} = \varepsilon_{vp}$) is expressed as follows:

$$\frac{d\varepsilon_{vp}}{dt} = \left(\frac{\sigma_{eq}}{\sigma_o}\right)^n \frac{\partial \sigma_{eq}}{\partial \boldsymbol{\sigma}} \tag{5.29}$$

Perzyna-type elastic-visco-plastic laws are used for representing nonlinear rate dependency involving plasticity (Perzyna 1966).

$$\frac{d\varepsilon_{vp}}{dt} = \lambda \mathbf{s} \tag{5.30}$$

where λ is the proportionality coefficient, and it is interpreted as a fluidity coefficient. This parameter is obtained from uniaxial creep experiments as:

$$\lambda = \frac{\dot{\varepsilon}_c}{\sigma} \tag{5.31}$$

5.3.2.2.2 Elasto-visco-plasticity

Another approach was proposed by Aydan and Nawrocki (1998), in which the material behavior is visco-elastic before yielding and becomes visco-plastic after yielding. The incremental elasto-visco-plastic constitutive law takes the following form:

$$d\boldsymbol{\sigma} = \mathbf{D}^{rp} d\boldsymbol{\varepsilon} + \mathbf{C}^{rp} d\dot{\boldsymbol{\varepsilon}} \tag{5.32}$$

where

$$\mathbf{D}^{rp} = \mathbf{D}^{r} - \frac{\mathbf{D}^{r} \dfrac{\partial G}{\partial \boldsymbol{\sigma}} \otimes \dfrac{\partial F}{\partial \boldsymbol{\sigma}} \mathbf{D}^{r}}{h_{rp} + \dfrac{\partial F}{\partial \boldsymbol{\sigma}} \cdot \left(\mathbf{D}^{r} \dfrac{\partial G}{\partial \boldsymbol{\sigma}}\right) + \dfrac{\partial F}{\partial \boldsymbol{\sigma}} \cdot \left(\mathbf{C}^{r} \dfrac{\partial \dot{G}}{\partial \boldsymbol{\sigma}}\right)} \tag{5.33}$$

$$\mathbf{C}^{rp} = \mathbf{C}^{r} - \frac{\mathbf{C}^{r} \dfrac{\partial \dot{G}}{\partial \boldsymbol{\sigma}} \otimes \dfrac{\partial F}{\partial \boldsymbol{\sigma}} \mathbf{C}^{r}}{h_{rp} + \dfrac{\partial F}{\partial \boldsymbol{\sigma}} \cdot \left(\mathbf{D}^{r} \dfrac{\partial G}{\partial \boldsymbol{\sigma}}\right) + \dfrac{\partial F}{\partial \boldsymbol{\sigma}} \cdot \left(\mathbf{C}^{r} \dfrac{\partial \dot{G}}{\partial \boldsymbol{\sigma}}\right)} \tag{5.34}$$

Figure 5.2 illustrates the elasto-visco-plastic model for one-dimensional response.

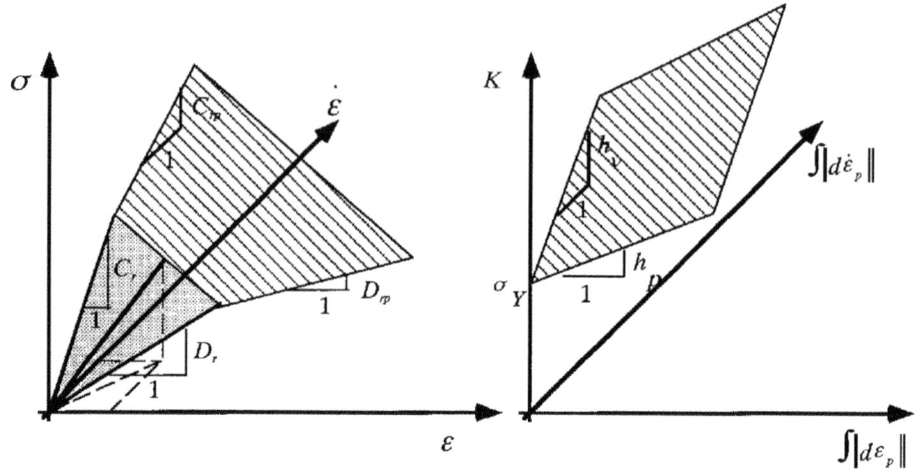

FIGURE 5.2 Elasto-visco-plastic model for one-dimensional response.

5.4 YIELD/FAILURE CRITERIA

Non-linear behavior requires the existence of yield functions. These yield functions are also called failure functions at the ultimate state when rocks rupture. It is common to use the Mohr–Coulomb yield criterion given by:

$$\tau = c + \sigma_n \tan\phi \quad \text{or} \quad \sigma_1 = \sigma_c + q\sigma_3 \tag{5.35}$$

where c, ϕ and σ_c are cohesion, friction angle and uniaxial compressive strength. σ_c and q are related to the cohesion and friction angle in the following form:

$$\sigma_c = \frac{2c\cos\phi}{1-\sin\phi} \quad \text{and} \quad q = \frac{1+\sin\phi}{1-\sin\phi} \tag{5.36}$$

As the intermediate principal stress is indeterminate in the Mohr–Coulomb criterion, and there is a corner-effect problem during the determination of the incremental elasto-plasticity tensor, the use of the Drucker–Prager criterion (1952) is quite common in numerical analyses, which is given by:

$$\alpha I_1 + \sqrt{J_2} = k \tag{5.37}$$

where

$$I_1 = \sigma_I + \sigma_{II} + \sigma_{III}\,;\, J_2 = \frac{1}{6}\left((\sigma_I - \sigma_{II})^2 + (\sigma_{II} - \sigma_{III})^2 + (\sigma_{III} - \sigma_I)^2\right)$$

Nevertheless, it is possible to relate the Drucker–Prager yield criterion with the Mohr–Coulomb yield criterion. On the π-plane, if the inner corners of the Mohr–Coulomb yield surface are assumed to coincide the Drucker–Prager yield criterion, the following relations may be derived (Figure 5.3):

$$\alpha = \frac{2\sin\phi}{\sqrt{3}(3+\sin\phi)} \quad k = \frac{6c\cos\phi}{\sqrt{3}(3+\sin\phi)} \tag{5.38}$$

where c, ϕ are cohesion and friction angle, respectively.

In Rock Mechanics, one of the recent yield criteria is Hoek–Brown's criterion (1980, 1997), which is written as:

$$\sigma_1 = \sigma_3 + \sqrt{m\sigma_c\sigma_3 + s\sigma_c^2} \tag{5.39}$$

where m and s are some coefficients. While the value of s is 1 for intact rock, the values of m and s change when they are used for rock mass.

Aydan (1995) proposed a yield function for thermo-plasticity yielding of rock as given by

$$\sigma_1 = \sigma_3 + [S_\infty - (S_\infty - \sigma_c)e^{-b_1\sigma_3}]e^{-b_2T} \tag{5.40}$$

where S_∞ is the ultimate deviatoric strength, while coefficients b_1, b_2 are empirical constants.

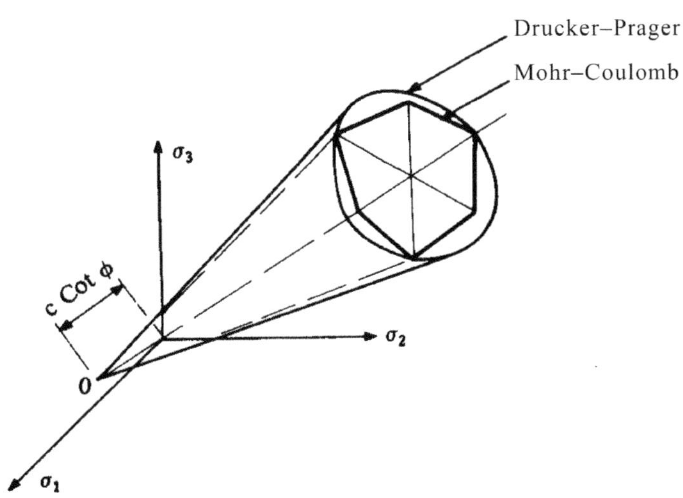

FIGURE 5.3 Yield criteria in principal stress space (from Owen-Hinton 1980).

Interested readers can consult the textbooks by Aydan (2020a, 2021c) for the comparison of yield (failure) criteria to actual experimental results involving igneous, metamorphic and sedimentary rocks.

Under very high confining pressure up to 1.5 GPa, Figure 5.4 shows an application of Eq. (5.40) to experiments on eclogite reported by Ryabinin et al. (1971) for the isothermal condition (Aydan 2021c), together with the fitted failure criteria of Mohr–Coulomb and Hoek–Brown. Although more detailed discussions are given in the next section, the Hoek–Brown criterion cannot evaluate experimental results from the brittle to ductile regime at all, and it is entirely an invalid yield criterion.

Hirth and Tullis (1994) reported the results of triaxial experiments on quartz aggregates under very high confining pressures, which is almost six times its uniaxial compressive strength. The best fit to the

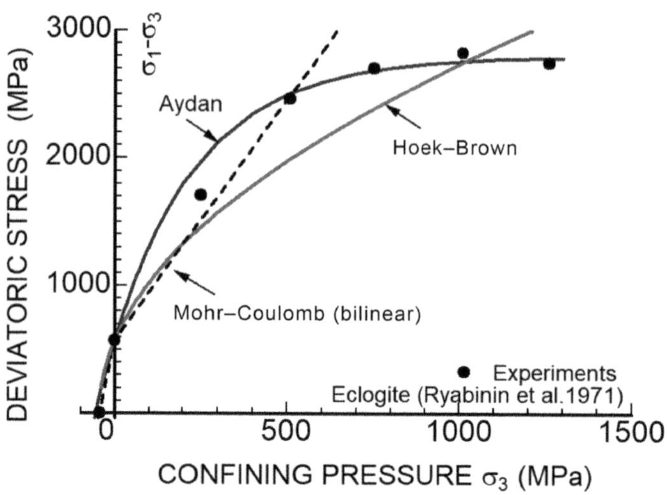

FIGURE 5.4 Comparison of various yield criteria with experimental results (Aydan 2021c).

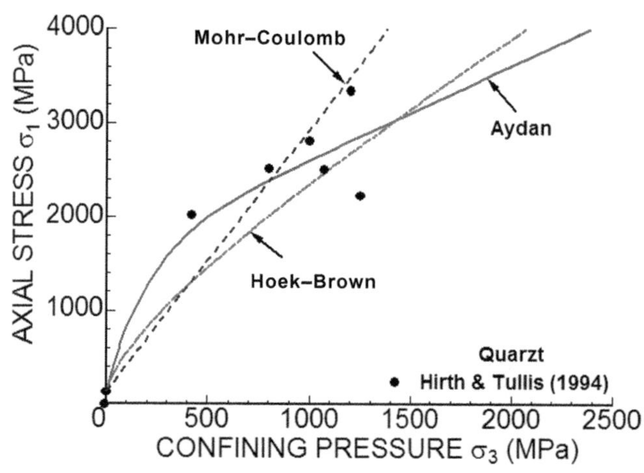

FIGURE 5.5 Comparisons of yield criteria for experimental results on quartz.

experimental results is obtained for Aydan's criterion as seen in Figure 5.5. Up to a confining pressure of 1000 MPa, the estimation by Mohr–Coulomb criterion is better than that by the Hoek–Brown criterion. Again, a very high discrepancy is observed among the estimations by the Hoek–Brown criterion and experimental results.

In geomechanics, there is almost no yield (failure) criterion incorporating the effect of temperature on yield (failure) properties of rocks, although there are some experimental researches (Hirth and Tullis 1994). Aydan's criterion is the only criterion known incorporating temperature, and it was used to study the stress state of the Earth (Aydan 1995). Figure 5.6 shows the experimental results for three different

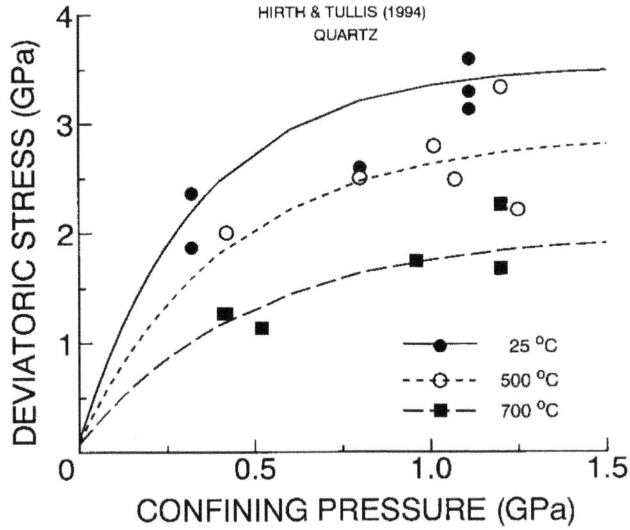

FIGURE 5.6 Experimental results of Hirth and Tullis (1994) on quartz for three different ambient temperatures.

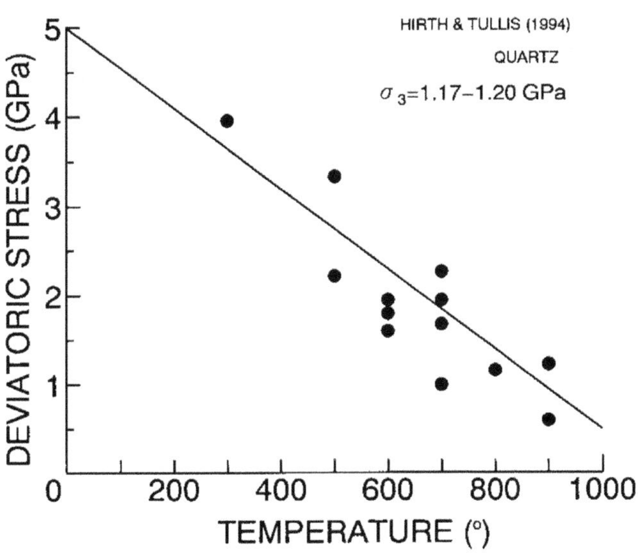

FIGURE 5.7 Reduction of deviatoric strength of quartz as a function of temperature for a confining pressure of 1.17–1.2 GPa.

values of ambient temperature reported by Hirth and Tullis (1994), while Figure 5.7 shows the reduction of strength with temperature for a given confining pressure of 1.17–1.2 GPa.

Aydan's yield (failure) criterion is applied to the experimental results shown in Figures 5.8 and 5.9, and the results are shown in Figure 5.8.

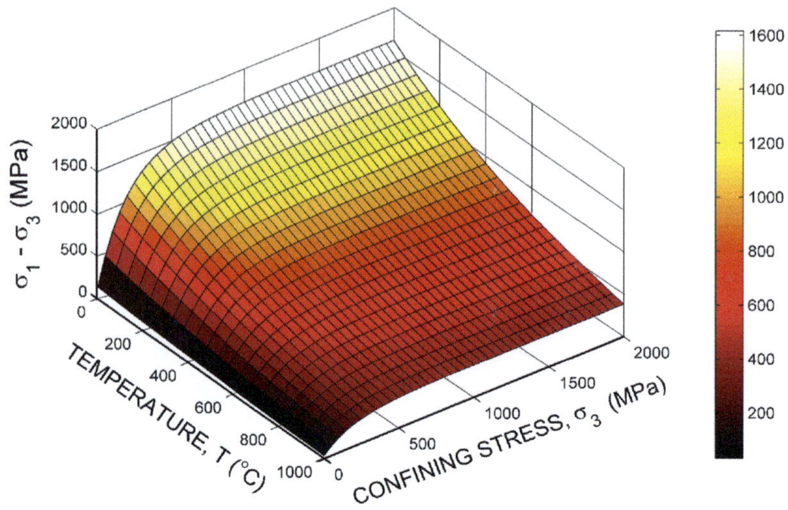

FIGURE 5.8 Three-dimensional representation of Aydan's failure criterion for experimental results of Hirth and Tullis (1994).

5.5 TESTING FOR EVALUATING PHYSICO-MECHANICAL PROPERTIES OF ROCKS

The parameters of constitutive laws are determined from tests in laboratory and/or in situ. Specifically, the following parameters are determined for rock and discontinuities for simplified conditions: cohesion, tensile strength, uniaxial compressive strength, friction angle, elastic modulus, Poisson's ratio for rocks. Figure 5.9 illustrates several testing techniques for rocks: uniaxial compression, triaxial compression, direct and indirect (Brazilian) tensile strength, bending test. There are also different techniques to determine the tensile strength of rocks, as illustrated in Figure 5.10. However, the tensile strength obtained from these techniques is generally different from one another. The cohesion and friction angle of rocks can also be determined from direct shear tests. Figure 5.11 shows direct shear testing equipment in the University of the Ryukyus, which is capable of carrying out shear tests under static, creep and dynamic loading conditions.

5.6 TESTING FOR SHEAR CHARACTERISTICS OF ROCK DISCONTINUITIES

As stated previously, discontinuities are one of major characteristics of rock masses in nature and found in various forms such as cracks, joints, schistosity, bedding and flow planes, and faults. The relevance of the static shear properties (e.g., shear behavior, deformability, strength, time-dependency etc.) of discontinuities is well-established in Rock Mechanics, as it plays a key role in many rock

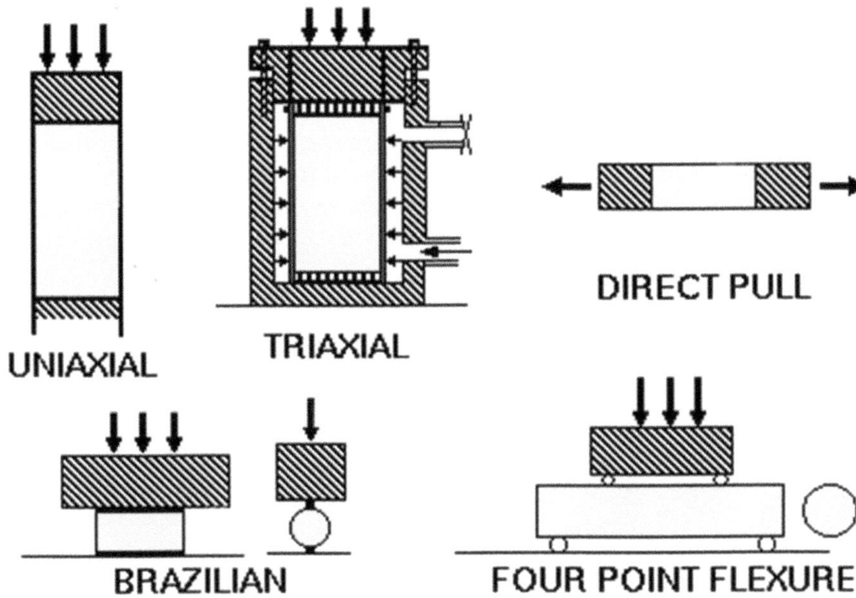

FIGURE 5.9 Various equipment configurations in laboratory testing (from Aydan 2020a).

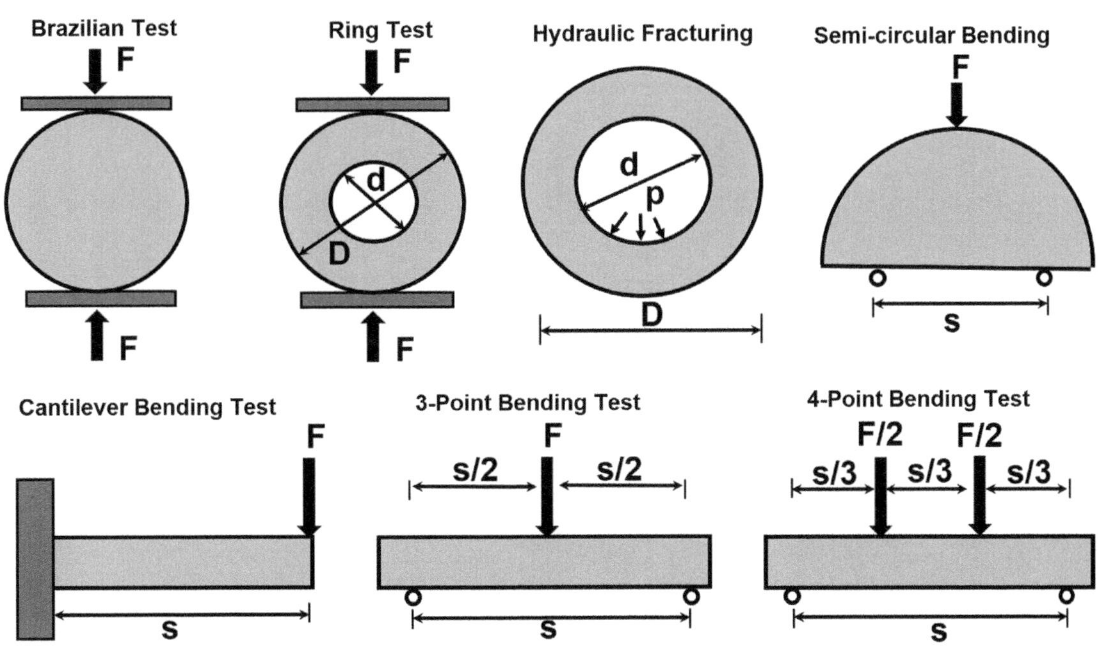

FIGURE 5.10 Several configurations for determining tensile strength (from Aydan 2022c).

engineering projects, such as natural or excavated slopes, dam foundations or underground excavations. The evaluation of the shear behavior of rock discontinuities under static, creep, cyclic and dynamic loading conditions is also necessary in relation to shearing, faulting, post-failure slip, induced vibrations (Figure 5.12).

The device shown in Figure 5.11 can also be used for shear testing of rock discontinuities under different loading conditions under high-normal stress levels. However, the devices for dynamic shear testing should also satisfy the requirements related to the intended purpose of utilization of experimental results. Some experiments are utilized to evaluate the dynamic shear strength characteristics of rocks using shock/impact-type loading (Figure 5.13).

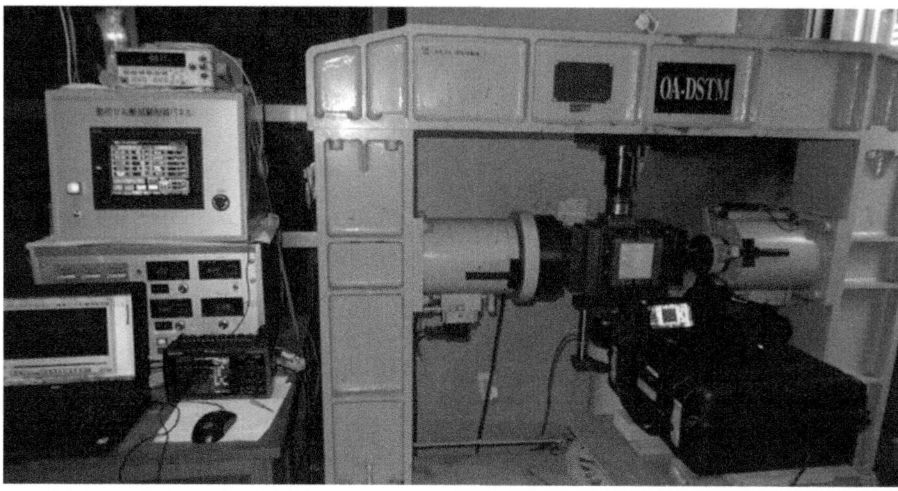

FIGURE 5.11 Dynamic shear testing device (OA-DSTM) (from Aydan 2021c).

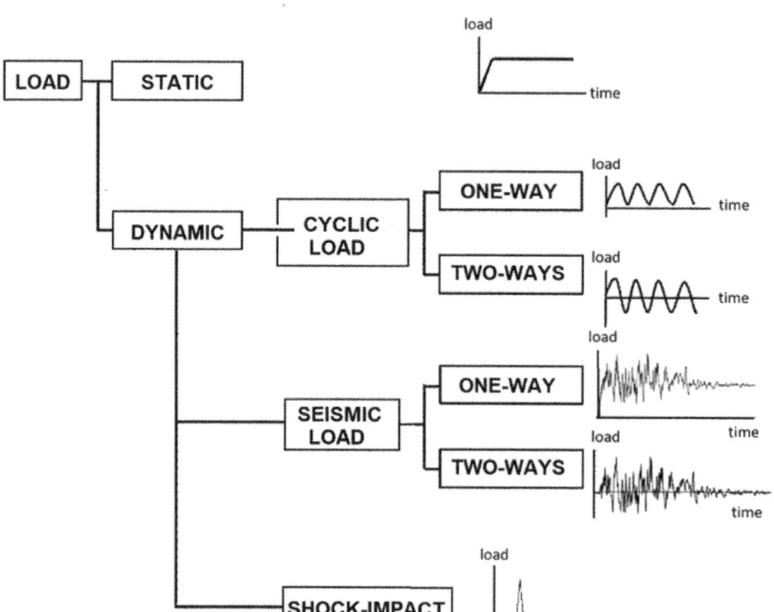

FIGURE 5.12 Classification of load patterns for shear testing of discontinuities.

Shaking tables, ranging from large installations to small devices, may also be used to test the shear properties of rock discontinuities, together with the utilization of appropriate equipment (Figure 5.14). In addition, the frictional properties of rock discontinuities under low-normal stress levels can be determined from tilt-test and stick-slip tests, as shown in Figures 5.15 and 5.16 (from Aydan et al. 2023).

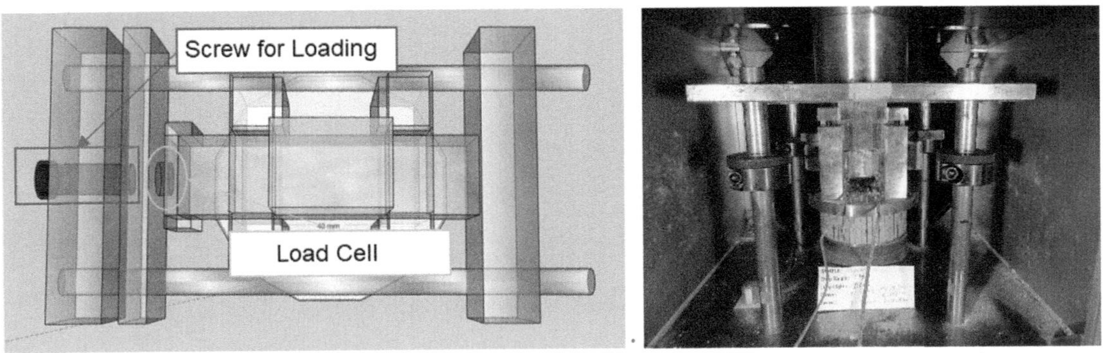

FIGURE 5.13 Double shear test device for rocks and discontinuities under shock load.

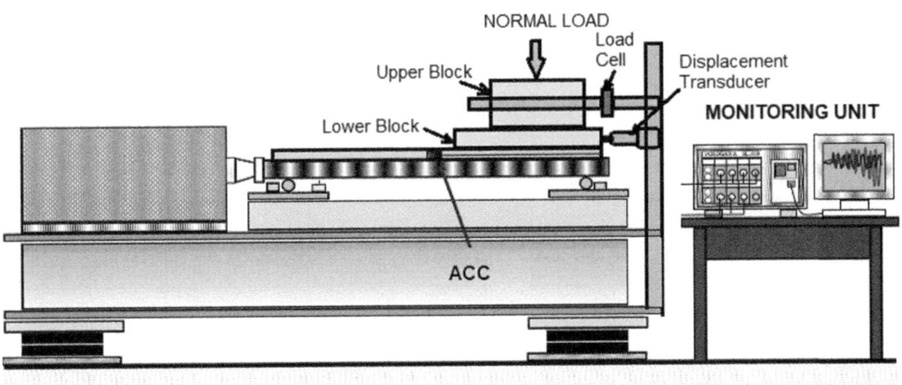

FIGURE 5.14 Dynamic shear testing device utilizing a shaking table (from Aydan 2019).

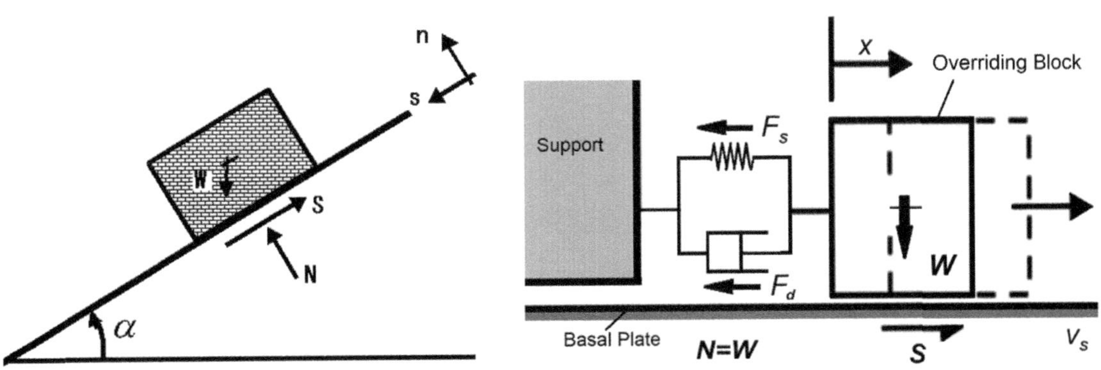

(a) Principle of tilt-test

(b) Principle of a stick-slip test

FIGURE 5.15 Principle of (a) tilt and (b) stick-slip test.

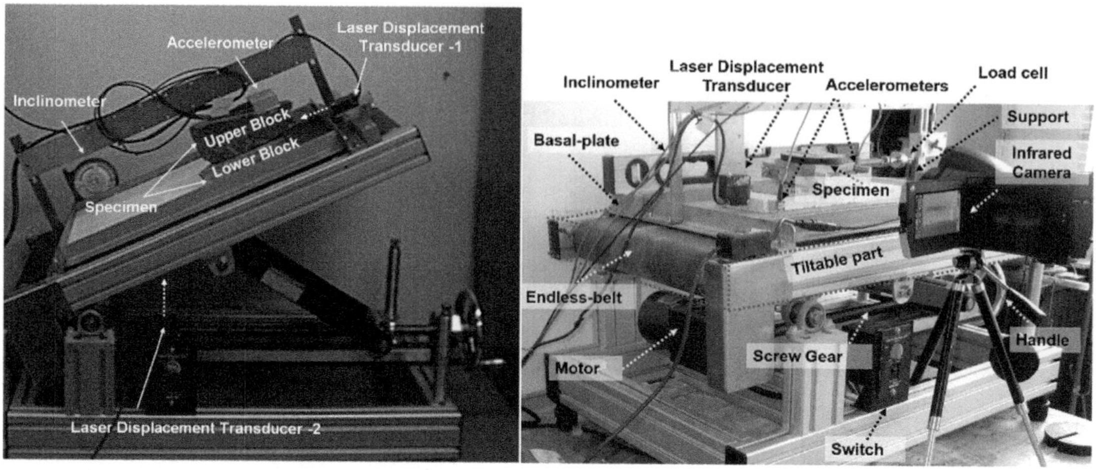

(a) Tilt-test (b) Stick-slip test

FIGURE 5.16 Views of (a) tilt-test and (b) a stick-slip test (from Aydan et al. 2023).

Stress State of Terrestrial Planetary Bodies

6

6.1 HYDROSTATIC APPROACH

The in situ stress state is one of the most important items when rock mechanics and rock engineering are to be utilized on the Moon and other terrestrial planets. Particularly, the engineering design requires data on the in situ stress state as well as quake-related topics. Figure 6.1 shows the density, gravity and hydrostatic pressure distributions with depth in Venus and Mars, together with those of Earth reported by Steinberger et al. (2010), which models the planetary objects as hydrostatic spherical bodies. It should be noted that the simulations do not involve the deformability properties of materials constituting the interior structures of the planets and are not taken into account in hydrostatic simulations.

The in situ stress measurements and estimations by one or several empirical, analytical and numerical methods can be used (e.g., Aydan 1993, 1995, 2000a, 2016, 2019; Amadei and Stephanson 1997). However, proper knowledge and information on the characteristics of materials constituting terrestrial planetary bodies are necessary. For example, the elastic constants obtained from wave velocities, density and gravitational distributions may be used for stress states in the terrestrial planetary bodies.

6.2 THERMO-ELASTO-PLASTIC CONSTITUTIVE MODELS FOR ROCKS OF TERRESTRIAL PLANETS

Many experiments on Earth rocks under high pressure have indicated that rocks exhibit elastic-strain-hardening behavior. For example, Hirth and Tullis [1994] reported the results of triaxial experiments on quartz aggregates under very high confining pressures, almost six times its uniaxial compressive strength. Furthermore, they also performed experiments under a temperature regime up to 900°C. Many yield criteria for the initiation and evolution of plastic behavior of rocks are presented in Chapter 5. However, Aydan's criterion is the only yield criterion that takes into account the effect of temperature on the triaxial strength of rocks. Aydan's yield function for the thermo-plasticity yielding of rock is given by Eq. (5.40) in Chapter 5. This yield criterion is used in the numerical simulations presented in the upcoming subsections.

DOI: 10.1201/9781003643852-6

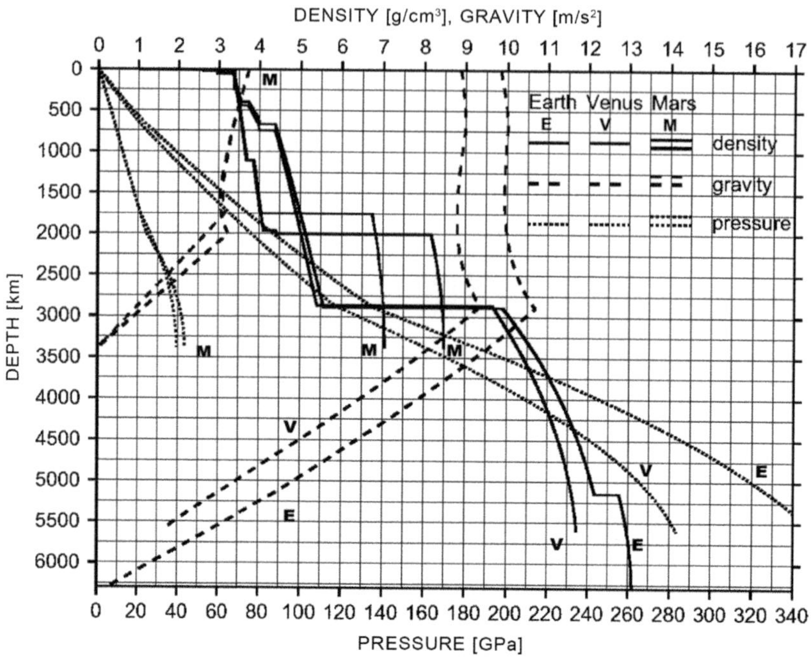

FIGURE 6.1 Comparison of the density, gravity and hydrostatic pressure distributions with depth in Venus and Mars, together with those of Earth (from Steinberger et al. 2010).

6.3 SPHERICAL SYMMETRIC FINITE ELEMENT MODEL FOR TERRESTRIAL PLANETS

The governing equation for a spherical symmetric body can be written as follows (Aydan 1993, 1995, 2016, 2020a, 2021b):

$$\frac{\partial \sigma_r}{\partial r} + 2\frac{\sigma_r - \sigma_\theta}{r} = \rho g \tag{6.1}$$

where ρ and g are density and gravitational acceleration, which may vary with depth. The constitutive law for elastic behavior is written as:

$$\begin{Bmatrix} \sigma_r \\ \sigma_\theta \end{Bmatrix} = \begin{bmatrix} 2\mu + \lambda & 2\lambda \\ \lambda & 2(\mu + \lambda) \end{bmatrix} \begin{Bmatrix} \varepsilon_r \\ \varepsilon_\theta \end{Bmatrix} \tag{6.2}$$

It should be noted that Lame's constants in the preceding equation can vary with depth. Taking variations on δu, the following integral form of that expression is obtained as:

$$\int_\Omega \left(\frac{\partial \sigma_r}{\partial r} + 2\frac{\sigma_r - \sigma_\theta}{r} \right) \delta u d\Omega = \int_\Omega \rho g \delta u d\Omega \tag{6.3}$$

where $d\Omega = 4\pi r^2 dr$. With the use of Gauss divergence theorem and the following conditions:

$$\delta u = 0 \quad \text{on} \quad \Gamma_u, \quad \text{and} \quad \hat{t} = 0 \quad \text{on} \quad \Gamma_t$$

The weak form of Eq. (6.3) takes the following form:

$$\int_{\Omega} (\sigma_r \delta \varepsilon_r + 2\sigma_\theta \delta \varepsilon_\theta) d\Omega = \int_{\Omega} \rho g \delta u d\Omega \tag{6.4}$$

Let us assume that the displacement in a given element is approximated by the following expression:

$$u = [N]\{U\} \tag{6.5}$$

where [N] is the shape function and $\{U\}$ is the nodal displacement vector. Using this approximate form and the constitutive law (6.4), the following finite element form is obtained for a typical finite element:

$$[K]\{U\} = \{F\} \tag{6.6}$$

where

$$[K] = \int_{\Omega_e} [B]^T [D][B] d\Omega,; \quad \{F\} = \int_{\Omega_e} \rho g \ [N]^T \ d\Omega$$

If a linear-type shape function is chosen:

$$N^i = \frac{r_j - r}{L}, \quad N^j = \frac{r - r_i}{L}, \quad L = r_j - r_i \tag{6.7}$$

Stiffness matrix $[K]$ and load vector $\{F\}$ given in Eq. (6.7) are specifically obtained as follows:

$$[K] = \frac{4\pi}{L^2} \begin{bmatrix} K_{11} & K_{12} \\ K_{12} & K_{22} \end{bmatrix}, \quad \{F\} = \frac{\rho g \pi}{3} \begin{Bmatrix} r_j^3 + r_j^2 r_i + r_j r_i^2 - 3r_i^3 \\ 3r_j^3 - r_j^2 r_i - r_j r_i^2 - r_i^3 \end{Bmatrix} \tag{6.8}$$

where

$$K_{11} = \frac{D_1 + 2D_2 + D_1^*}{3}(r_j^3 - r_i^3) - (D_2 + D_1^*) r_j (r_j^2 - r_i^2) + D_1^* r_j^2 (r_j - r_i)$$

$$K_{22} = \frac{D_1 + 2D_2 + D_1^*}{3}(r_j^3 - r_i^3) - (D_2 + D_1^*) r_i (r_j^2 - r_i^2) + D_1^* r_i^2 (r_j - r_i)$$

$$K_{12} = K_{21} = -\frac{D_1 + 2D_2 + D_1^*}{3}(r_j^3 - r_i^3) + (D_2 + D_1^*)\frac{r_j + r_i}{2}(r_j^2 - r_i^2) - D_1^* r_i r_j (r_j - r_i)$$

$$D_1 = 2\mu + \lambda, \quad D_2 = 2\lambda \quad D_1^* = 4(\mu + \lambda)$$

6.4 STRESS ANALYSES OF TERRESTRIAL PLANETS

6.4.1 Earth

In analyses reported herein, it was assumed that the stress–strain response of rocks constituting Earth exhibit an elastic-perfectly plastic behavior (Aydan 1995, 2020a). The elastic constants of rocks were taken from a report by Anderson and Hart (1976). The initial stiffness technique was chosen as the iteration technique to deal with the plastic behavior of rocks in finite element analyses (Owen and Hinton 1980). The temperature distribution of Earth was input as known by using a distribution reported in a textbook (p. 248) by Fowler (1990) (Figure 6.2).

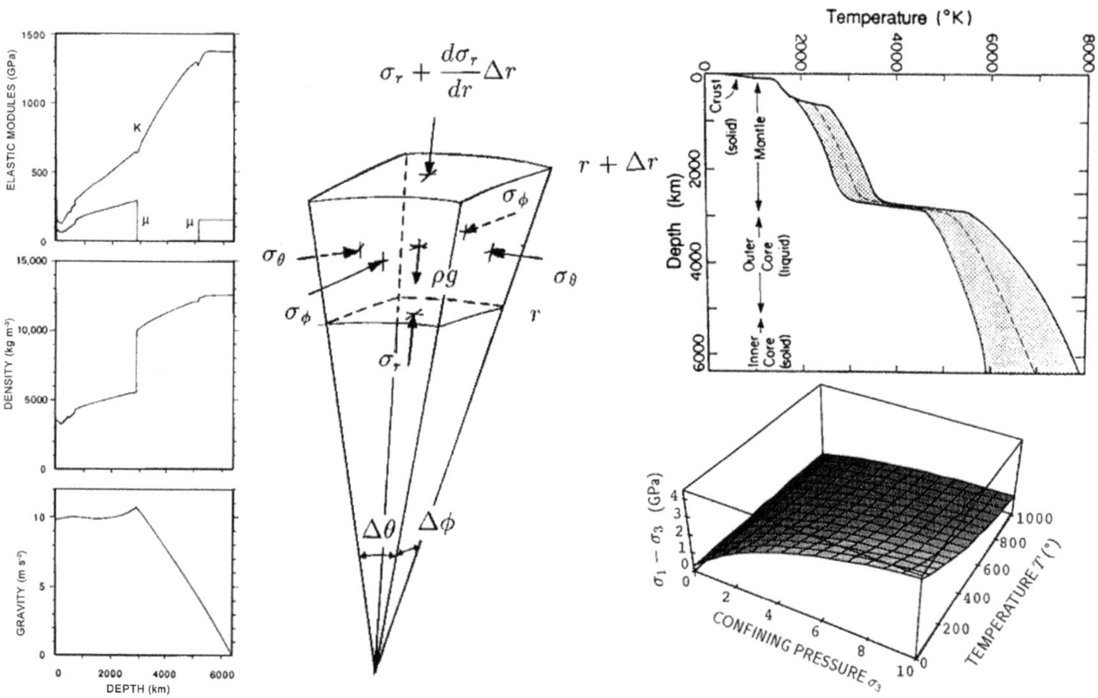

FIGURE 6.2 Illustration of spherical model of Earth and material properties (arranged from Aydan 1994 and Fowler 1990).

A series of case studies given here were carried out by Aydan (1995) (Figure 6.3):

CASE 1: Earth was in liquid state (hydrostatic).

CASE 2: The crust and mantle were elastic solids, and the core was in the liquid state (non-hydrostatic).

CASE 3: The crust and mantle were elasto-plastic solids, and the core was in the liquid state (isothermic: room temperature).

CASE 4: The crust and mantle were thermo-elasto-plastic solids, and the core was in the liquid state (non-isothermic).

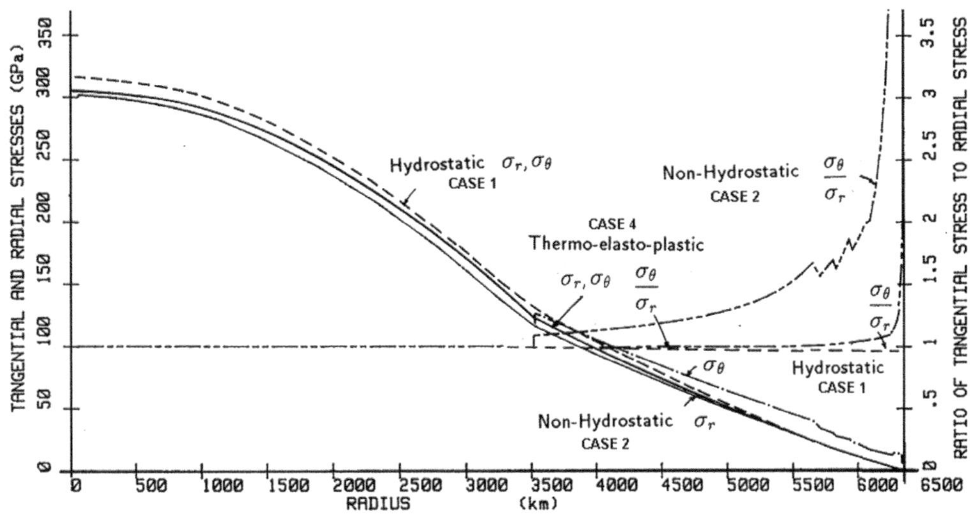

FIGURE 6.3 Stress state of Earth (from Aydan 1995).

Parameters b_1, b_2, S_o, S_∞ of the yield criterion were chosen as 0.2 GPa^{-1}, 0.0014 $^{\circ}C^{-1}$, 0 MPa, 5 GPa, respectively. Figure 6.4 shows the distributions of radial and tangential stresses and the ratio of tangential stress to radial stress in the lithosphere for each case. The elasto-plastic analyses showed that the whole crust and mantle became plastic. Large tangential stresses seen in CASE 2 were dissipated when the plastic behavior of the crust and mantle was considered. The tangential stresses dissipated in the mantle increased both radial and tangential stresses in the core. The consideration of the thermo-elasto-plastic behavior of the mantle and the crust of the Earth further decreased the deviatoric stresses in the mantle and in the crust.

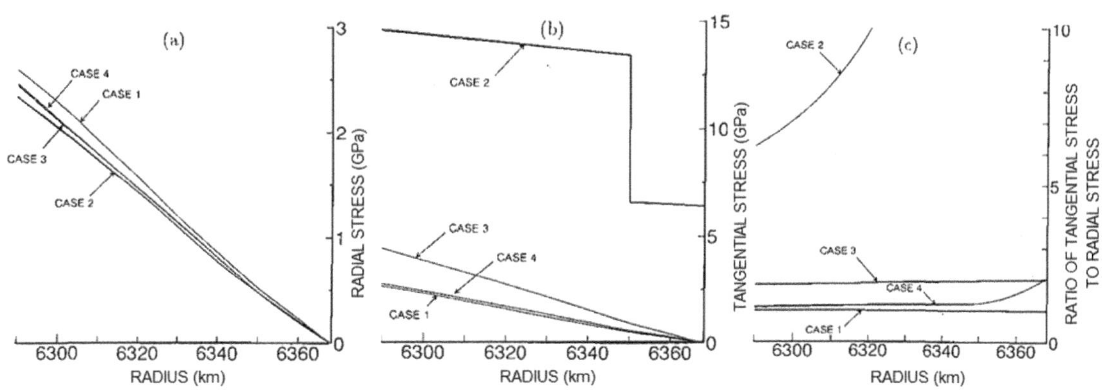

FIGURE 6.4 Computed radial and tangential stresses in lithosphere.

The lateral stress coefficient, which has a value of infinity at the ground surface for CASE 2, also decreased in magnitude when the plastic behavior was considered. In the next series of parametric studies, the uniaxial compressive strength parameter S_o was varied from 0 MPa to 40 MPa by an increment of 20 MPa, while the other parameters were kept the same as those in the previous thermo-elasto-plastic analysis. The computed distributions of radial and tangential stresses and the ratio of tangential stress to radial stress, together with in situ observations (Aydan and Paşamehmetoğlu 1994) and predictions by Shimada (1993) for a depth of 4 km in the Earth's crust, are plotted in Figure 6.5. The radial stress (vertical stress) almost coincided with each other (Figure 6.5[a]). They are almost equal to the vertical stress calculated from $\rho g H$ (Shimada 1993) and a good fit to in situ measurements.

The tangential stresses (horizontal stress) for each value of S_o fit fairly well to in situ measurements for a depth of 1 km from the ground surface (Figure 6.5[b]). However, finite element computations overestimate the tangential stresses at depths greater than 1 km.

The ratio of tangential stress to radial stress for each value of S_o provide upper bounds for in situ measurements (Figure 6.5[c]). The discrepancy between computed and measured values may be eliminated if a thermo-elasto-visco-plastic constitutive law is employed in finite element analysis.

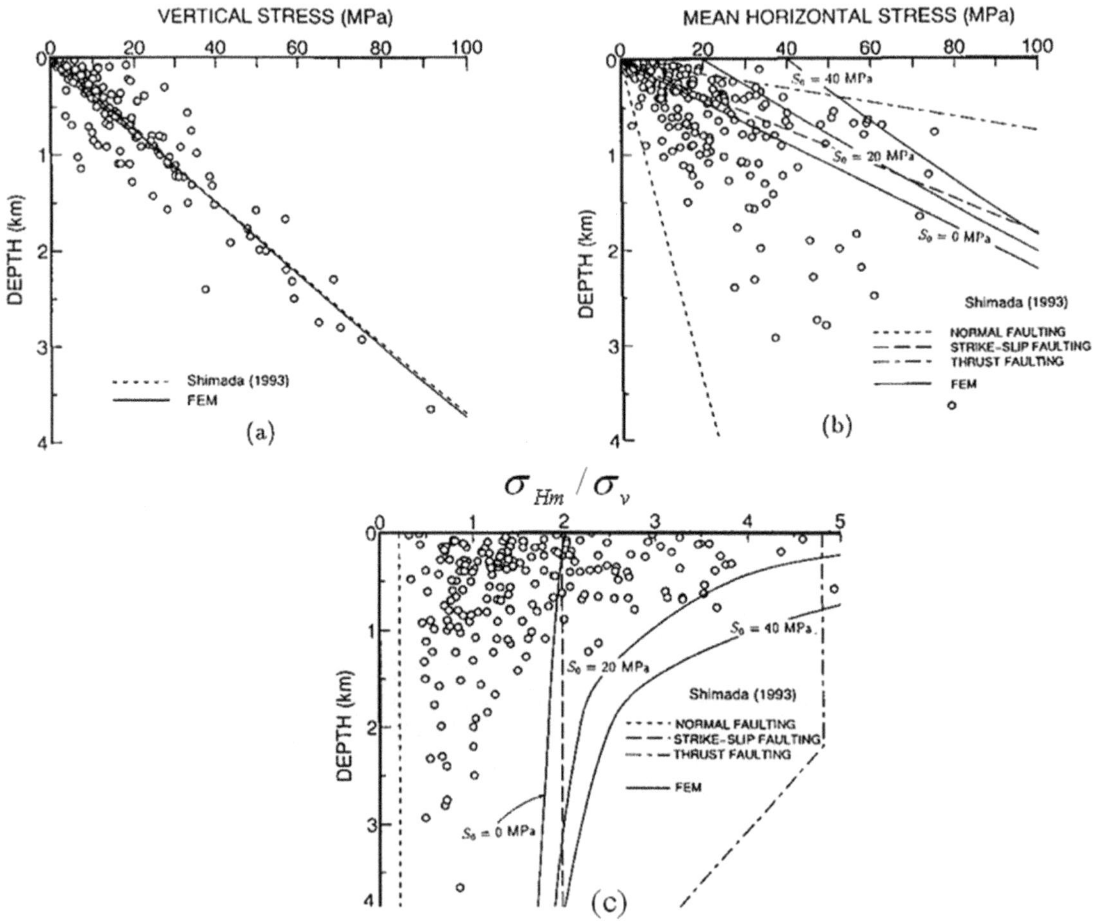

FIGURE 6.5 Comparison of computed and measured in situ stresses (from Aydan and Kawamoto 1994).

6.4.2 Moon

Next, the stress state of the Moon is analyzed as it is the most extensively explored planetary body by humankind. The results are summarized in subsection 3.2 and Figure 6.6. The elastic properties were computed from relations presented in Chapter 3 and those shown in Figure 6.6. The temperature

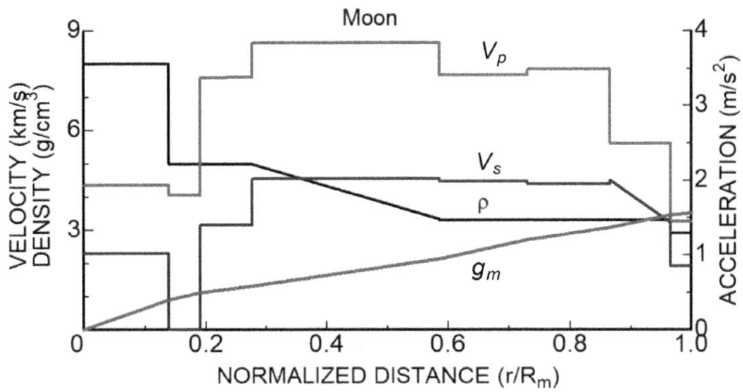

FIGURE 6.6 Distributions of gravity, density, elastic wave velocities through the Moon.

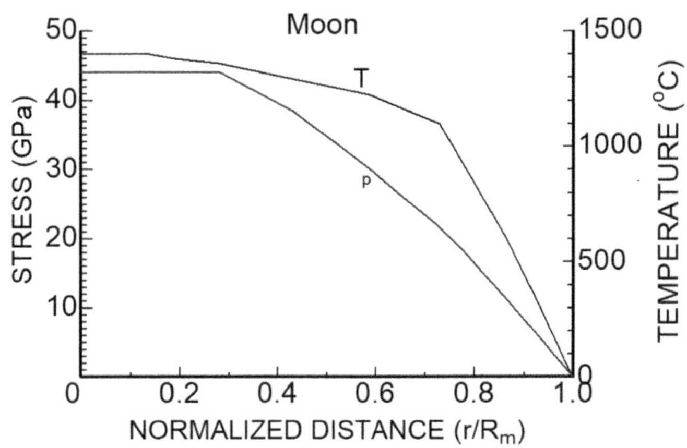

FIGURE 6.7 Distribution of temperature and pressure through the Moon.

distribution is assumed as shown in Figure 6.7 on the basis of the past pioneering studies (e.g., Toksöz 1974, 1975; Toksöz et al. 1972, 1974, 1977; Dainthy et al. 1976; Weber et al. 2011).

Using the spherical planetary model together with constitutive laws described in Chapter 5, the stress state of the Moon is computed for different models for the following four conditions:

CASE 1: Hydrostatic situation
CASE 2: Elastic properties with the consideration of solid and fluid states
CASE 3: Elasto-plastic properties with the consideration of solid and fluid states
CASE 4: Thermo-elasto-plastic properties for solid parts

Parameters of the yield function are the same as those shown in Figure 5.20 and computations for Earth.

Figure 6.8 shows the stress states computed for CASE 1, CASE 2, CASE 4. A discontinuity in tangential stress is noted at a radius of 330 km, which corresponds to the liquid and solid interface for the model shown in Figure 3.12(b). Yielding occurs partially in the zones bound between radius 330 km and radius 376 km and between radius 1564 km and the surface of the Moon. The zone between radius

376 km and radius 1564 km remains elastic. In other words, except for the core and upper 174 km, the behavior is elastic for a great part of the Moon on the basis of computational results for the assumed conditions.

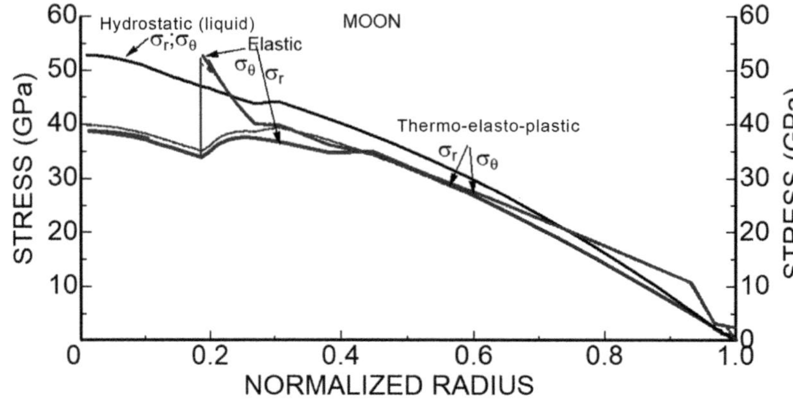

FIGURE 6.8 Stress states for the Moon computed for different conditions.

A close-up view of the stress state for the conditions of CASE 1, CASE 2, CASE 4 is shown in Figure 6.9. The tangential stress is 244 MPa for the elastic state, and it is reduced to 0 MPa upon yielding.

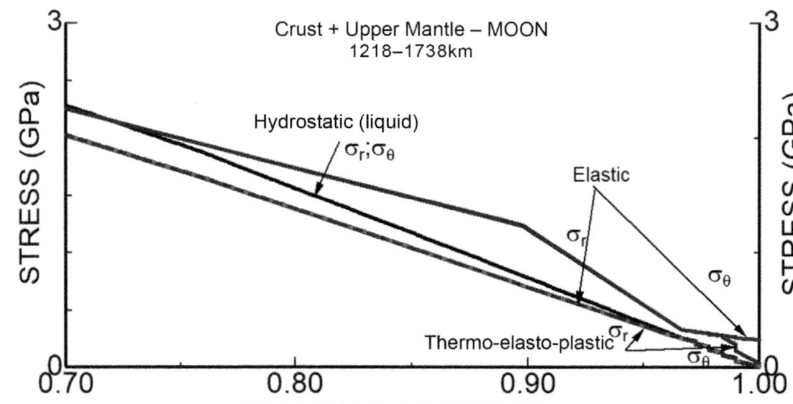

FIGURE 6.9 Stress state in lithosphere for the conditions of CASE 1, CASE 2, CASE 4.

Next the ratio of tangential (horizontal) stress to radial (vertical) is plotted and shown in Figure 6.10 for the conditions of CASE 1, CASE 2, CASE 4. While the ratio goes to infinity for the elastics case (CASE 2), the ratio is limited to 2.46 for CASE 4. This implies that the tangential (horizontal) stress is the maximum principal stress while the radial stress is the least principal stress.

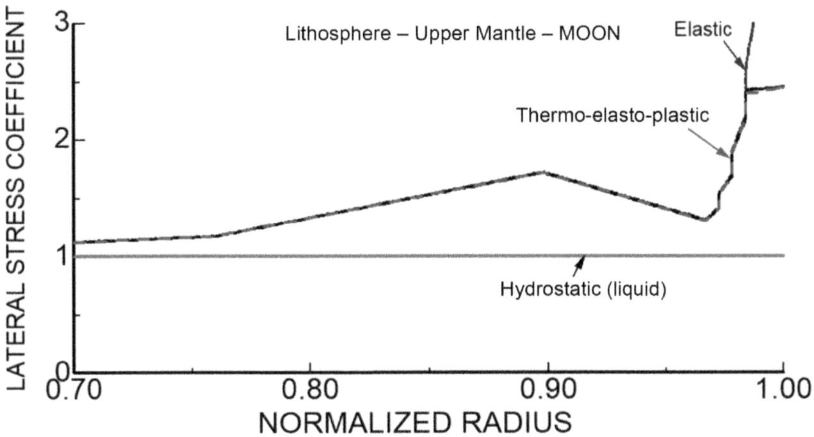

FIGURE 6.10 Ratio of tangential stress to radial stress in the Moon for various conditions.

6.4.3 Mars

In analyses, it was assumed that the stress–strain response of rocks constituting Mars exhibit an elastic-perfectly plastic behavior. The elastic constants of rocks were computed from the distributions of density, P-wave and S-wave velocities, as shown in Figure 6.11. The temperature distribution of the Earth shown in Figure 6.12 was input, using a distribution proposed by Zharkov and Gudkova (2005).

The series of case studies given next were carried out:

CASE 1: Mars was the in liquid state (hydrostatic).
CASE 2: The crust and mantle were elastic solids, and the core was in the liquid state (elastic).

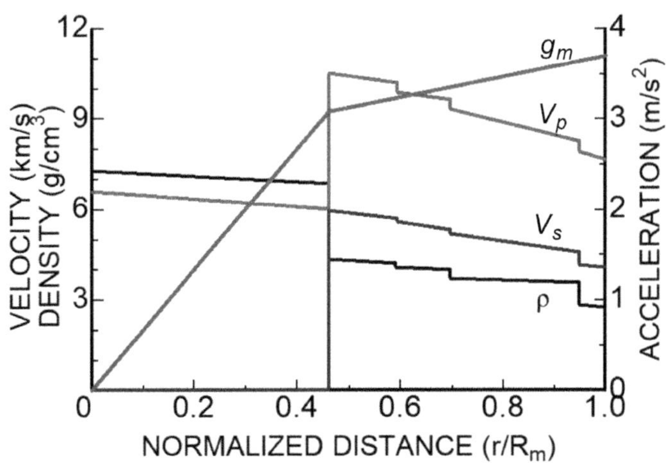

FIGURE 6.11 Distributions of gravity, density, elastic wave velocities through Mars.

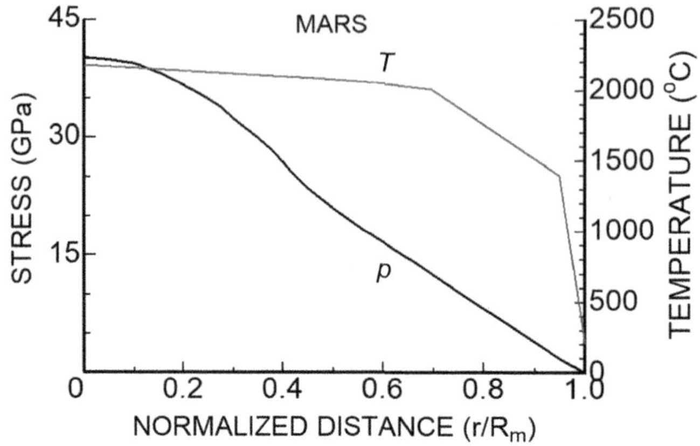

FIGURE 6.12 Distribution of temperature and pressure through Mars.

CASE 3: The crust and mantle were elasto-plastic solids, and the core was in the liquid state (isothermic: elasto-plastic).

CASE 4: The crust and mantle were thermo-elasto-plastic solids, and the core was in the liquid state (non-isothermic).

Parameters b_1, b_2, S_o, S_∞ of the yield criterion were chosen as 0.5 GPa^{-1}, 0.0033 $^{\circ}C^{-1}$, 0 MPa, 3 GPa, respectively, and they are the same as those used in previous analyses. Computed distributions of radial and tangential stresses in Mars's lithosphere are shown in Figure 6.13. Liquid state (CASE 1) and the elastic state (CASE 2) provide some bounds for radial and tangential stresses. If the mantle and crust of Mars are assumed to behave elastically together with the assumption that the core is in the liquid state, the tangential stress in the mantle and crust is higher than the radial stress. It is interesting to note that the value of the tangential stress is about 3.0 GPa at the ground surface. This value is almost the half of the tangential stress for Earth behaving elastically, as reported by Aydan (1993, 1995, 2016) and shown in Figures 6.3 and 6.4. As is well-known, the uniaxial compressive strength of the strongest rock on Earth is about 0.6 MPa. It is natural to expect that the mantle and crust of Mars should exhibit an elasto-plastic behavior under high tangential stress distribution.

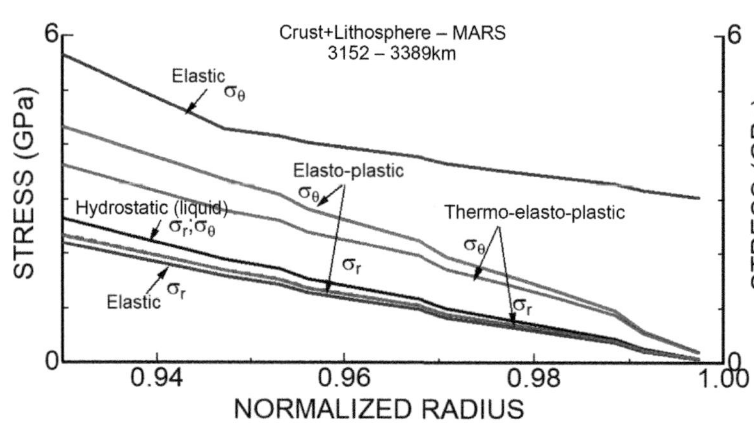

FIGURE 6.13 Computed radial and tangential stresses in the Mars lithosphere.

If mantle and crust is assumed to behave in an elasto-plastic manner, the high tangential stress for elastic behavior is greatly reduced, and the whole mantle and crust become plastic. As a result, the radial and tangential stress in the core were increased. If the temperature distribution estimated by Zharkov and Gudkova (2005) is utilized and the yield strength of materials constituting the mantle and crust of Mars obeys the yield criterion of Aydan (1995) with chosen material properties, the tangential stress in the mantle and crust is further reduced while the radial stress is increased. In addition, the hydrostatic radial and tangential stresses in the core are increased. In other words, the consideration of the thermo-elasto plastic behavior of the mantle and the crust of the Mars further decreased the deviatoric stresses in the mantle and in the crust. These results are fundamentally similar to those estimated for Earth by Aydan (1993, 1995, 2016) and are also shown in subsection 6.4.1.

As mentioned previously, the elasto-plastic analyses showed that the whole crust and mantle became plastic. Large tangential stresses seen in CASE 2 were dissipated when the plastic behavior of the crust and mantle was considered. The radial stress (vertical stress) almost coincided with each other up to a depth of 30 km from the surface. They are almost equal to the vertical stress calculated from $\rho g H$. The lateral stress coefficient shown in Figure 6.14, which has a value of infinity at the ground surface for CASE 2, also decreased in magnitude when the plastic behavior was considered. Lateral stress coefficient is roughly 2.34 for thermo-plastic behavior for a crust without uniaxial compressive strength.

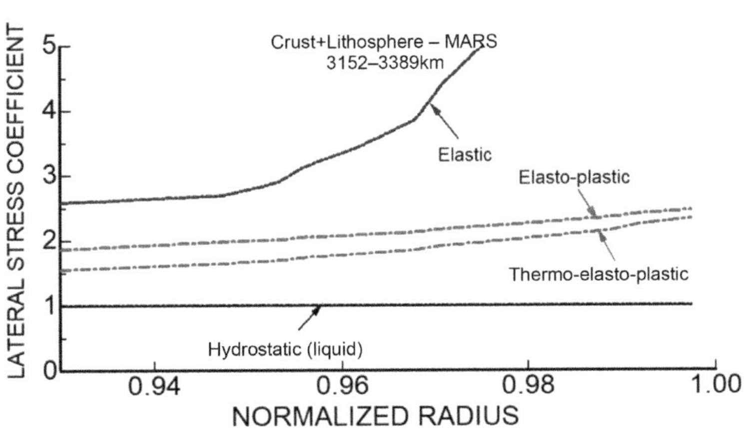

FIGURE 6.14 Distribution of lateral stress coefficient in the Mars lithosphere.

6.4.4 Venus

In the analyses for Venus, the elastic constants of rocks were computed from the distributions of density and P- and S-wave velocities are shown in Figure 6.15 on the bases of findings by pioneering researchers (e.g., Amorim and Gudkova 2023; Aitta 2012; Gudkova and Zharkov 2020). The temperature distribution of Venus shown in Figure 6.16 was input using a distribution proposed by Amorim and Gudkova (2023).

Similar to previous analyses, a series of case studies were carried out:

CASE 1: Venus was in the liquid state (hydrostatic).
CASE 2: The crust and mantle were elastic solids, and the core was in the liquid state (elastic).

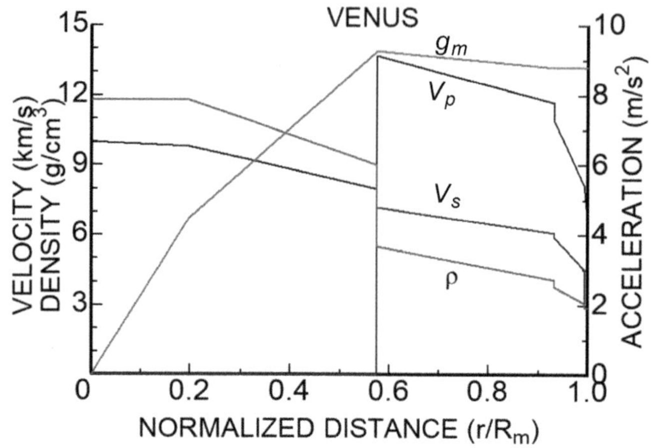

FIGURE 6.15 Distributions of gravity, density, elastic wave velocities through Venus.

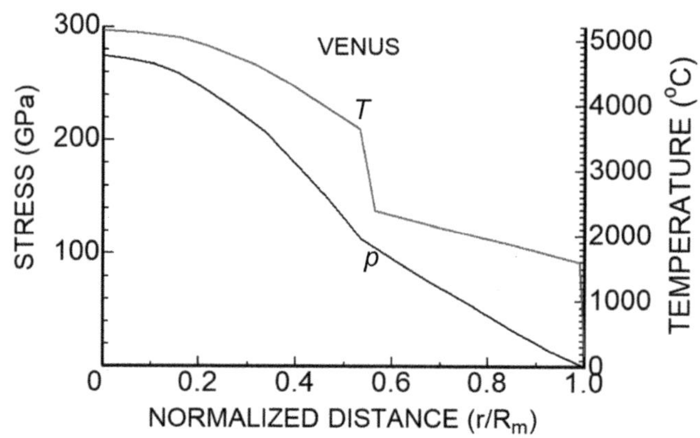

FIGURE 6.16 Distribution of temperature and pressure through Venus.

CASE 3: The crust and mantle were elasto-plastic solids, and the core was in the liquid state (isothermic: elasto-plastic).

CASE 4: The crust and mantle were thermo-elasto-plastic solids, and the core was in the liquid state (non-isothermic).

Yield parameters for elasto-plastic and thermo-plastic behavior are the same as before.

Computed distributions of radial and tangential stresses in the lithosphere of Venus are shown in Figure 6.17. Liquid state (CASE 1) and elastic state (CASE 2) provide some bounds for radial and tangential stresses. The value of the tangential stress at the ground surface is about 4.0 GPa. This value is almost 60% of the tangential stress for Earth behaving elastically, as reported by Aydan (1993, 1995, 2016) and shown in subsection 6.4.1. This difference is likely due to smaller elastic constants obtained from slightly lower P-wave and S-wave velocities. It should be noted that there are no measured P-wave and S-wave velocities in Venus yet and that they have been just obtained in analogy to those of Earth. Nevertheless, it is natural to expect that the mantle and crust of Venus should exhibit an elasto-plastic behavior under high tangential stress distribution.

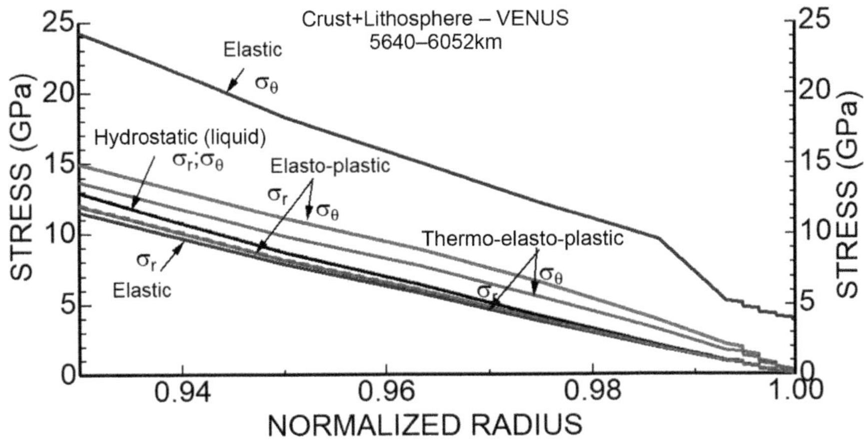

FIGURE 6.17 Computed radial and tangential stresses in the lithosphere of Venus.

If the mantle and crust are assumed to behave in an elasto-plastic manner, the high tangential stress for elastic behavior is greatly dissipated, and the whole mantle and crust of Venus become plastic. If the temperature distribution estimated by Amorim and Gudkova (2023) is utilized, and the yield strength of materials constituting the mantle and crust of Venus obeys the yield criterion of Aydan (1995) with chosen material properties, the tangential stress in the mantle and crust is further reduced while the radial stress is increased. In other words, the consideration of the thermo-elasto plastic behavior of the mantle and the crust of Venus further decreased the deviatoric stresses in the mantle and in the crust.

The lateral stress coefficients are shown in Figure 6.18 for four different conditions. It has a value of infinity at the ground surface for CASE 2, and it decreased in magnitude when the plastic behavior was considered. The lateral stress coefficient is roughly 2.28 for thermo-plastic behavior for a crust without uniaxial compressive strength. This value is slightly smaller as the temperature distribution is high in the lithosphere of Venus.

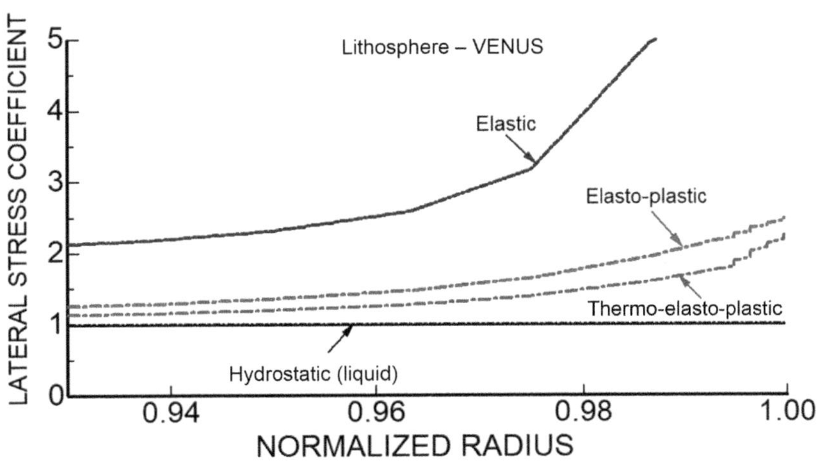

FIGURE 6.18 Distribution of lateral stress coefficient in the lithosphere of Venus.

6.4.5 Mercury

As for the analyses on the stress state of Mercury, the elastic constants of materials constituting Mercury were computed from the distributions of density and P-wave and S-wave velocities as shown in Figure 6.19. The temperature distribution of the Mercury shown in Figure 6.20 was input using a distribution proposed by several researchers (e.g., Solomon 1976; Lyttleton 1969; Michel et al. 2013; Tosi et al. 2013).

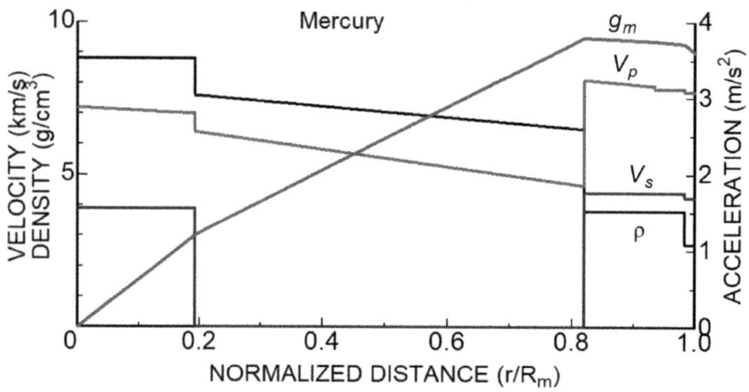

FIGURE 6.19 Distributions of gravity, density, elastic wave velocities through Mercury.

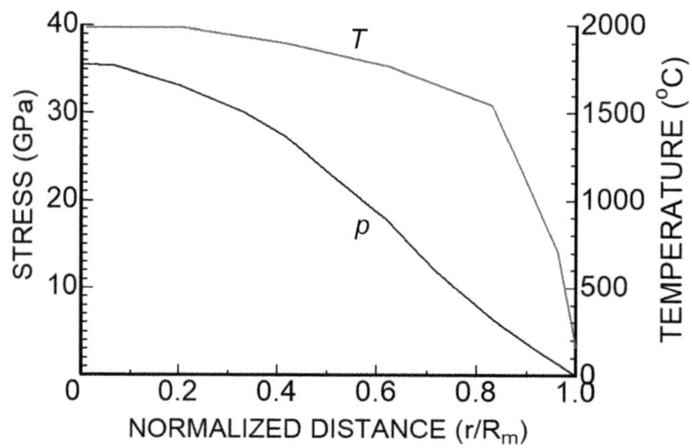

FIGURE 6.20 Distribution of temperature and pressure through Mercury.

As for the stress state of Mercury, the following series of case studies were carried out:

CASE 1: Mercury was in the liquid state (hydrostatic).
CASE 2: The crust and mantle were elastic solids, and the core was in the liquid state (elastic).

CASE 3: The crust and mantle were elasto-plastic solids, and the core was in the liquid state (isothermic: elasto-plastic).

CASE 4: The crust and mantle were thermo-elasto-plastic solids, and the core was in the liquid state (non-isothermic).

Yield parameters for elasto-plastic and thermo-plastic behavior are the same as used before. Computed distributions of radial and tangential stresses in Mercury lithosphere are shown in Figure 6.21. The liquid state (CASE 1) and elastic state (CASE 2) provide some bounds for radial and tangential stresses. The value of the tangential stress at the ground surface is about 3.1 GPa. This value is almost 47% of the tangential stress for Earth behaving elastically, as reported by Aydan (1993, 1995, 2016) and subsection 6.4.1. This difference is likely to be due to smaller elastic constants obtained from slightly lower P-wave and S-wave velocities, temperature distribution, and size of the planet. It should be also noted that there is no measured P-wave and S-wave velocities in Mercury yet and that they have just been obtained in analogy to those of Earth. Nevertheless, it is natural to expect that the mantle and crust of Mercury could not sustain such high tangential stress distribution.

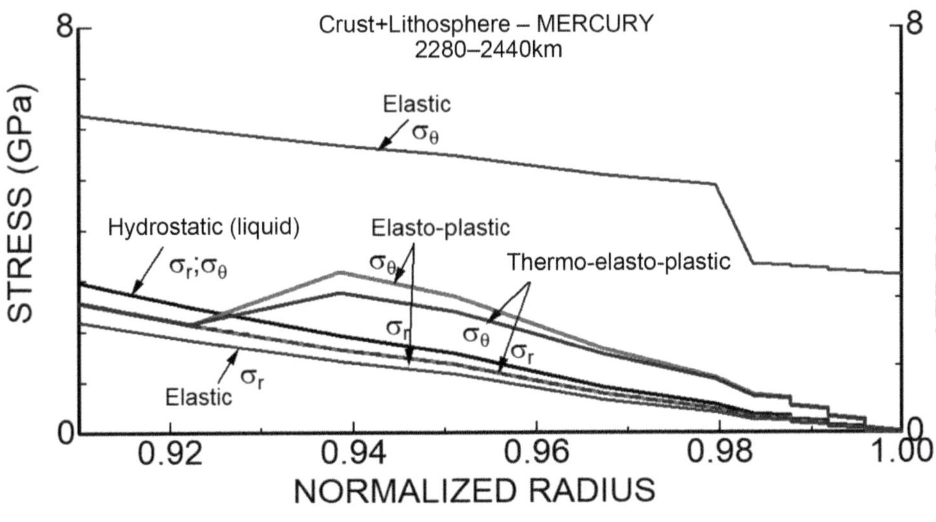

FIGURE 6.21 Computed radial and tangential stresses in the lithosphere of Mercury.

If the plastic behavior of the mantle and crust of Mercury is considered in analyses, the high tangential stress for elastic behavior is greatly reduced, and the whole mantle and crust of Mercury become plastic. If the temperature distribution estimated by researchers mentioned previously is utilized, and the yield strength of materials constituting the mantle and crust of Mercury obeys the yield criterion of Aydan (1995) with the chosen material properties, the tangential stress in the mantle and crust is further reduced while the radial stress is increased. In other words, the consideration of the thermo-elasto plastic behavior of the mantle and the crust of the Mercury further decreased the deviatoric stresses in the mantle and crust.

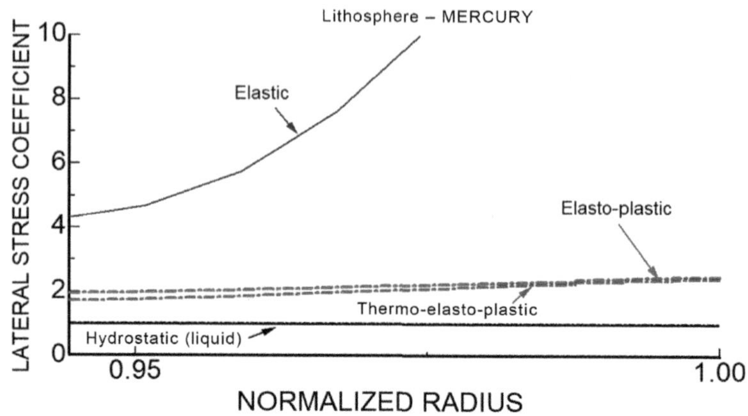

FIGURE 6.22 Distribution of lateral stress coefficient in the lithosphere of Mercury.

The lateral stress coefficients are shown in Figure 6.22 for four different conditions. It has a value of infinity at the ground surface for elastic behavior (CASE 2), and it decreased in magnitude when the plastic behavior was considered. Lateral stress coefficient is roughly 2.42 for thermo-plastic behavior for a crust without uniaxial compressive strength.

Seismicity in Terrestrial Planetary Bodies

7

Aydan (2017c, 2022a, 2022b, 2023b) stated that possible causes of quakes in planetary bodies are as follows:

1. Impacts by meteorites
2. Thermal contraction and expansion
3. Volcanic activity
4. Large-scale mass movements
5. Plate-tectonic-type movements
6. Daily rotation and annual solar motion

It should be noted that Earth is well-instrumented, while there is very limited instrumentation for seismic events on the Moon and Mars. In this chapter, quakes on Earth (Aydan 2023a, 2017b, 2002, 2003, 2006, 2007, 2009, 2012a; Aydan et al. 2005a, 2005b, 2007a, 2007b; Ulusay et al. 2001, 2003a, 2003b) are considered, and available knowledge and information are extended to other terrestrial bodies.

7.1 EARTHQUAKES

The main cause of the quakes on Earth is the stress changes in the Earth's crust and its temporary mechanical instability. The plate tectonic theory based on Wegener's theory on continental drift proposed in 1912 is often used to explain the occurrence of quakes of Earth, which are called "earthquakes."

It is often speculated that volcanic eruptions cause earthquakes or vice versa. Without any doubt, earthquakes are temporary instabilities in the Earth's crust, and the dissipation of accumulated mechanical energy during earthquakes occurs. Furthermore, there are some areas where creep-like deformation of the Earth's crust takes place. On the other hand, the intrusion of magma into the crust decreases the strength of the crust, altering the stress state in Earth's crust.

7.1.1 Determination of Occurrence Time

The motions induced by earthquakes can be recorded using strong motion equipment (acceleration, velocity) with short period and broadband seismographs. Particularly, broadband seismometers are used to record long-distance earthquakes. However, they may become saturated in the

DOI: 10.1201/9781003643852-7

close vicinity of an earthquake epicenter with large magnitudes. The occurrence time of an earthquake fundamentally utilizes the arrival time differences of P-wave and S-wave (see Figure 3.2 for an example record). Omori (1894) proposed the following relation for the distance between the hypocenter of an earthquake and time lag of wave arrivals at a given seismic observation point for P-wave and S-wave arrivals (Figure 7.1). Fundamentally, the earthquake occurrence time is unknown.

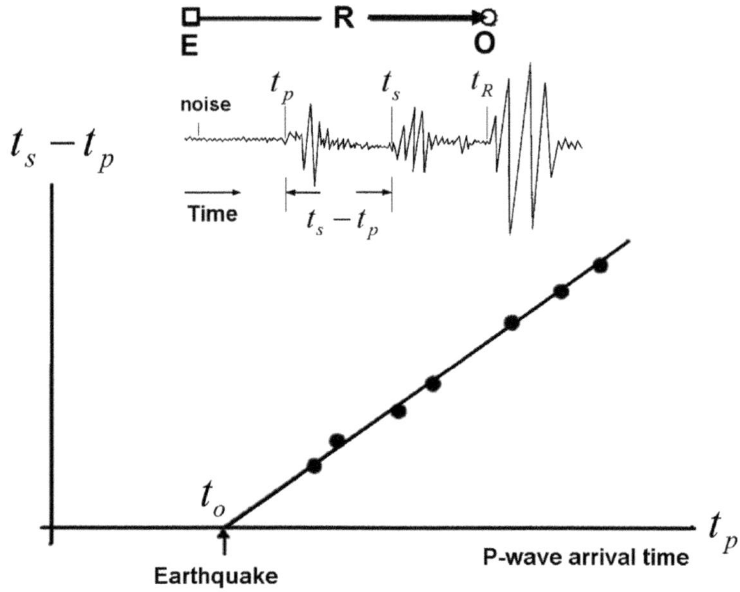

FIGURE 7.1 Arrival times of P-wave, S-wave and surface wave.

$$t_p - t_o = \frac{R}{V_p} \text{ and } t_s - t_o = \frac{R}{V_s} \tag{7.1}$$

Subtracting S-wave arrival time lag from the P-wave arrival time yields the following equation:

$$t_s - t_p = \left(\frac{1}{V_s} - \frac{1}{V_p} \right) R = \frac{R}{k} \tag{7.2}$$

where k, which is called Omori coefficient, takes the following form:

$$k = \frac{V_s V_p}{V_p - V_s} = \frac{1}{\frac{V_p}{V_s} - 1} V_p \tag{7.3}$$

V_p and V_s are P-wave and S-wave velocity of Earth's crust, which are not known. However, the following relation exists between these wave velocities in view of Eqs. (3.2) and (3.4):

$$\frac{V_s}{V_p} = \sqrt{\frac{1-2\nu}{2(1-\nu)}} \tag{7.4}$$

As the Poisson's ratio of Earth's crust ranges between 0.25 and 0.3, the following values for V_s / V_p ratio may be assumed:

$$\frac{V_s}{V_p} = 0.53 - 0.58 \tag{7.5}$$

Thus the value of the Omori coefficient would take the following value in terms of P-wave velocity as:

$$\frac{k}{V_p} = 0.724 - 0.887 \tag{7.6}$$

The average P-wave velocity of the Earth's crust ranges 6.5–7.2 km/s for upper-crust earthquakes and 7.5–8.2 km/s for lower-crust earthquakes.

7.1.2 Determination of Hypocenter and Epicenter

It is well-known that earthquakes are initiated at a certain depth. The initiation location of earthquakes is called the hypocenter. In this subsection, the determination of hypocenter for two-dimensional and three-dimensional situations is explained. The theory again utilizes the Omori's method.

7.1.2.1 Two-Dimensional (2D) Determination of Hypocenter and Epicenter

Let us consider two observation points denoted by A and B and an earthquake occurring along the straight line at a point denoted H, which will be assumed to be corresponding to the hypocenter, and its projection on the ground surface is E, which is called the epicenter (Figure 7.2).

Using Omori's formula, the hypo-central distance to the earthquake focus is given by:

$$R = V_p\left(t_p - t_o\right) \tag{7.7}$$

In this particular problem, besides the distance between points A and B, the hypocenter distances R_A, R_B can be computed from using the Omori's formula by assigning a value to Omori's coefficient. However, the results will not be generally affected by the value of Omori's coefficient. Thus the known parameters are:

$$R_A, R_B, L \tag{7.8}$$

On the other hand, the unknown parameters are

$$h, L_A, L_B \tag{7.9}$$

However, the unknown distances L_A, L_B are related to the total distance between points A and B though the following relation:

$$L_A + L_B = L \tag{7.10}$$

Let's introduce parameter α to relate the distances L_A, L_B to the total distance L as

$$\frac{L_B}{L} = (1-\alpha); \frac{L_A}{L} = \alpha \tag{7.11}$$

From the geometry one can easily determine parameter α as

$$\alpha = \frac{1}{2L^2}\left(R_A^2 + L^2 - R_B^2\right) \tag{7.12}$$

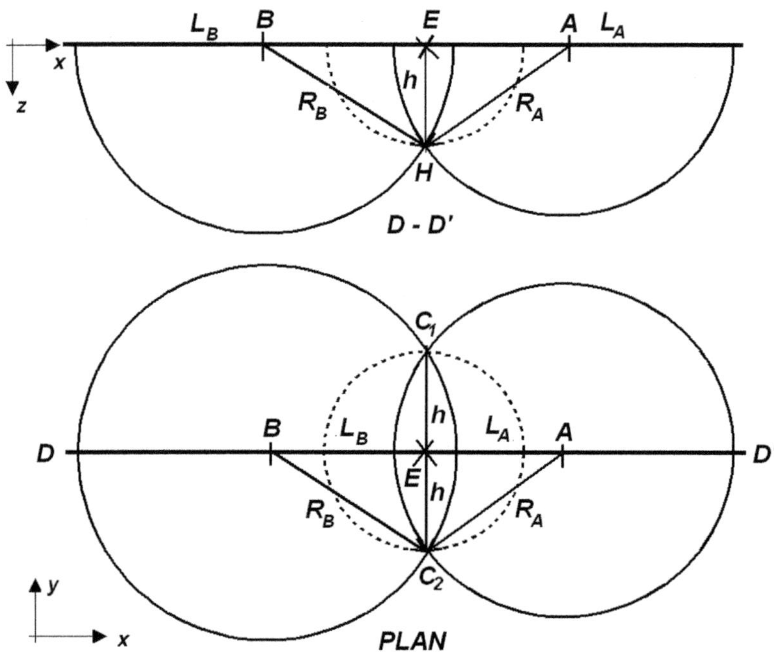

FIGURE 7.2 Hypocenter and epicenter for two-dimensional situation.

Once parameter α is determined, first L_A, L_B is determined, and then the focal depth is computed from one of following relations:

$$h = \sqrt{R_A^2 - L_A^2} \tag{7.13a}$$

or

$$h = \sqrt{R_B^2 - L_B^2} \tag{7.13b}$$

7.1.2.2 Three-Dimensional Determination of Hypocenter and Epicenter

Three-dimensional determination of hypocenter and epicenter can be carried out using the same procedure and/or graphically. Let us consider that there are three seismic stations with given coordinates and S-wave and P-wave arrival time differences, as shown in Figure 7.3. In the graphical method, the following procedure is adopted:

1. Using Omori's coefficient, circles with diameters obtained from Omori's method are drawn first, and then three common chords among circles are drawn.
2. The intersection of the three chords is denoted by E, which would correspond to the epicenter.
3. Then a semicircle having the shortest common chord length as its diameter is drawn, a straight line from E is drawn perpendicularly to the shortest chord, and the intersection point H is obtained. The length between E and H corresponds to the depth of the hypocenter.

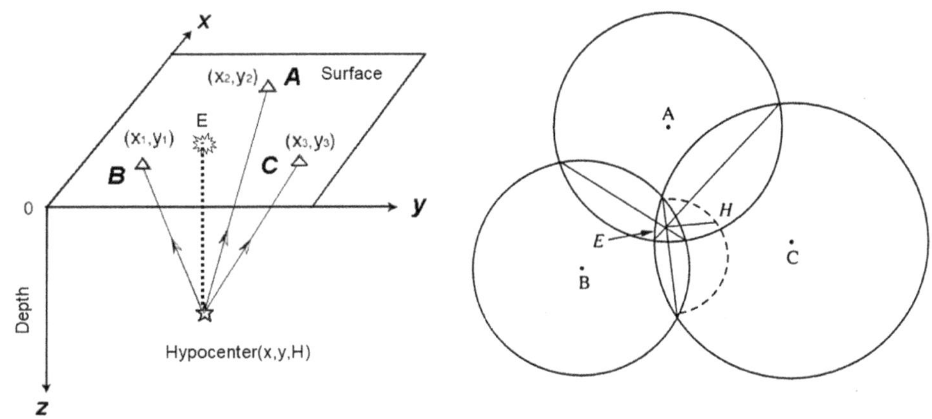

FIGURE 7.3 Illustration of three-dimensional hypocenter and epicenter determination procedure.

The concepts explained in the previous subsection and graphical method just explained can be combined. As the geometry becomes somewhat cumbersome, computational schemes would be necessary. For example, if the number of stations is assumed to be three, the radius for each station is obtained, then common chords for each pair of circles of stations, and the intersection point of the chords are obtained. Then the shortest chord length is searched, and the hypocenter is obtained. Figure 7.4 shows such a computational example of the procedure. The data used in this example are given in Table 7.1.

TABLE 7.1 Locations of Stations and Arrival Time Differences

STATION NAME	*X (km)*	*Y (km)*	Δt_{sp} *(s)*	$R = k\Delta t_{sp}$
A	50	50	6.25	50
B	120	50	3.125	25
C	90	90	3.75	30

$k = 8$ km/s

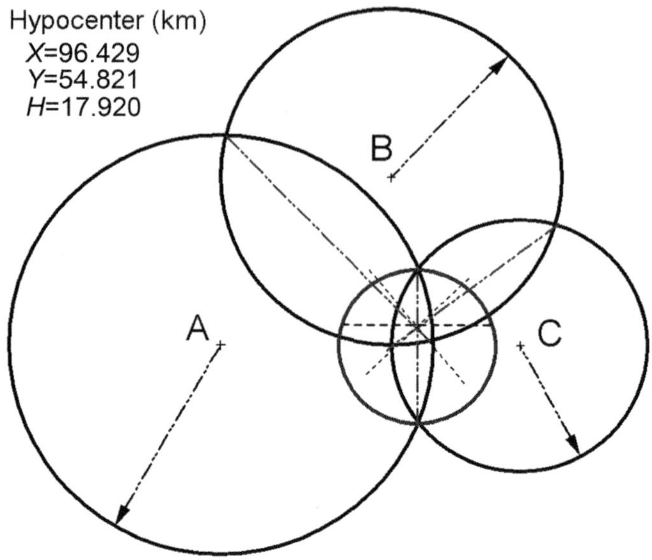

FIGURE 7.4 Sample solution.

In practice, there are many observation stations. The procedure is extended to observational data, while the fundamental procedure would remain the same. For observational data, the results would not be always the same, and some differences would occur. Furthermore, the velocity and seismic rays may be path dependent, and Earth is spherical. In general, such effects are considered. For this reason, some minimization procedure for differences is adopted, for which some commonly used software is available in the literature.

7.1.3 Determination of Magnitude

The energy E released during an earthquake can be obtained from following relation:

$$E = \int \Delta t \cdot u dS \tag{7.14}$$

where Δt is released traction on the fault surface, u is relative slip, S area of fault. If average values are used, this equation takes the following form:

$$E = \overline{t}S \cdot \overline{u} \tag{7.15}$$

Earthquake moment (M_0) is first defined by Aki (1966), and it is given in the following form (Figure 7.5):

$$M_o = \mu S\overline{u} \tag{7.16}$$

where μ is the shear modulus of the earth crust; S is the rupture area; $\overline{u}$ is average relative slip. The average relative slip is generally related to the maximum relative slip $u_{\max}$ as:

$$u_{max} = 2\bar{u} \tag{7.17}$$

Through field reconnaissance, GPS and DInSAR data can be used to obtain the relative slip of the causative earthquake, and the seismic records are used to obtain the seismic moment. It should be noted the seismic moment obtained from various seismological institutes would generally differ from one another. Hanks and Kanamori (1979) defined moment magnitude utilizing seismic moment, in the following form:

$$M_w = \frac{2}{3}\log_{10} M_0 - 10.7 \tag{7.18}$$

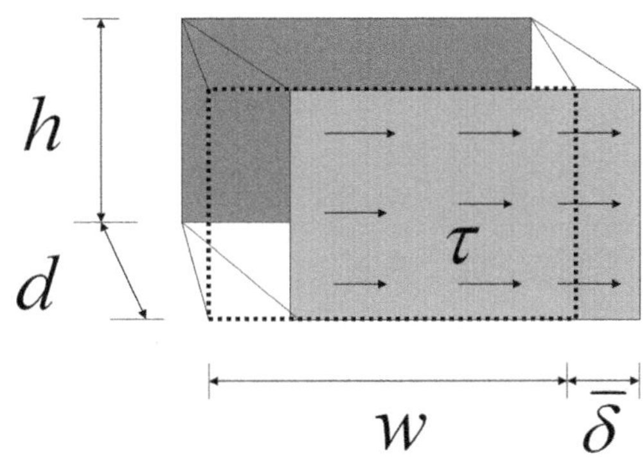

FIGURE 7.5 Illustration of seismic moment.

The concept of earthquake magnitude was first introduced by Richter (1935), and it is known as local earthquake magnitude (M_L) in modern times. Richter (1935) based his magnitude on seismographs measured by the Wood–Anderson torsion seismograph located at a distance of 100 km. Richter (1935) prepared a scale to determine the magnitude of the earthquake, as shown in Figure 7.6.

All earthquake magnitude scales utilize the maximum amplitude of a seismogram measured from the seismograph and its period, together with some correction factors counting local conditions and the distance:

$$M = \log_{10}\left(\frac{A}{T}\right) + q(\Delta, h) + a \tag{7.19}$$

where A is amplitude, T is period, $q(\Delta, h)$ is a correction factor depending upon the distance, and a is an empirical factor.

There are different magnitude scales of earthquakes depending upon the utilized wave type:

M_b: Body wave magnitude (P-wave or S-wave)
Ms: Surface wave magnitude (surface wave)
M_d: Duration magnitude (duration of seismogram for small earthquake)
M_L: Local magnitude (Richter type)
M_w: Moment magnitude

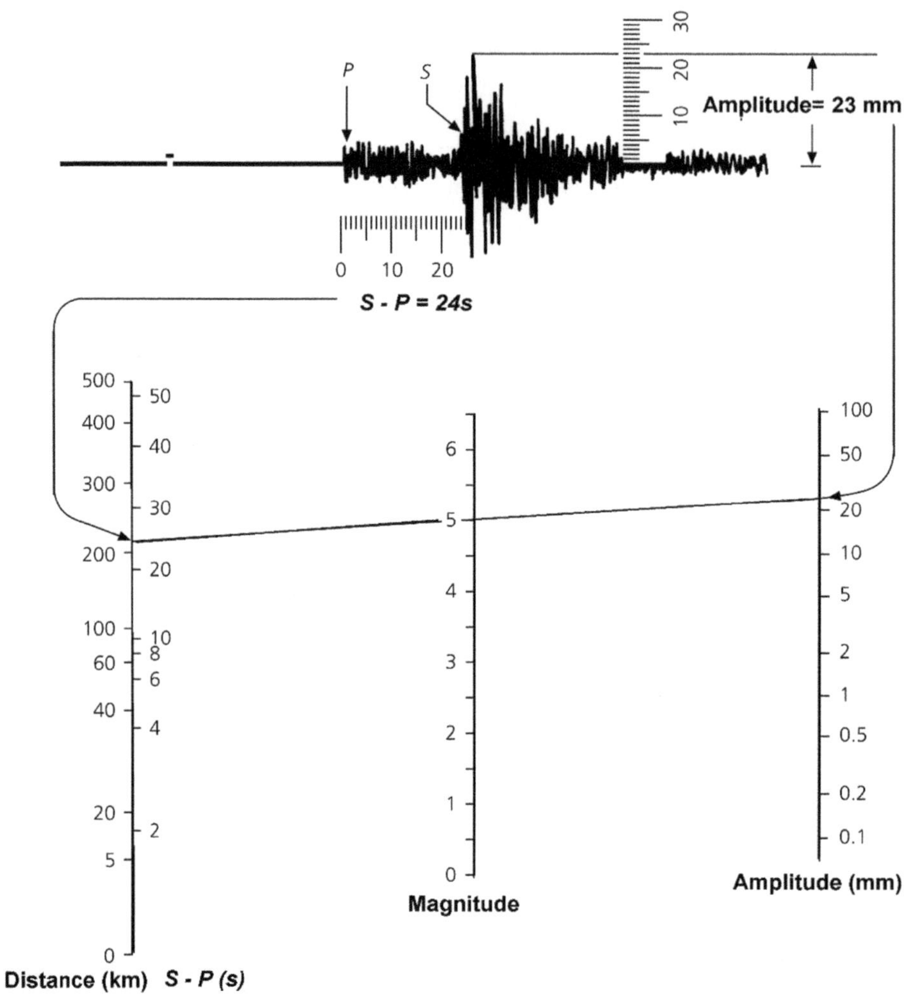

FIGURE 7.6 Manual chart of Richter to determine the magnitude of earthquake (from Richter 1935).

The correction factor generally depends upon the local conditions of seismographs and the distance from the hypocenter. When earthquakes are very large, seismograms may be saturated. On the other hand, deep earthquakes could not be determined by Richter type and surface wave–based seismographs.

7.1.4 Characteristics of Earthquake Faults

A fault is geologically defined as a discontinuity in geological medium along which a relative displacement took place. Faults are broadly classified into three big groups: normal faults, thrust (reverse) faults and strike-slip faults, as seen in Figure 7.7. The fault is geologically defined as active if a relative movement took place within a period less than 2 million years ago.

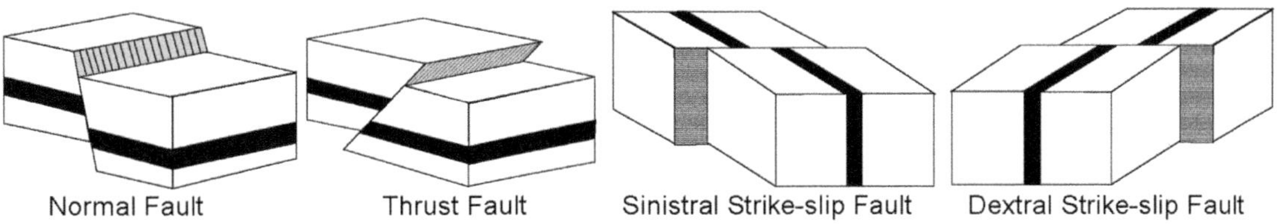

FIGURE 7.7 Fault types (from Aydan 2003, 2012a).

7.1.5 Characterization of Earthquakes from Fault Ruptures

Türkiye is one of the well-known earthquake-prone countries in the world, and most of her large earthquakes involve ground surface rupturing. The data from the past and present earthquakes of Turkey as well as other countries may be quite useful to establish and/or to revise empirical relations among the characteristics of earthquakes accompanying ground surface rupturing. The data compiled by the author are from the Turkish earthquake database (TEDBAS) developed by Aydan (1997) and additional inputs from recent earthquakes (Ambraseys 1988; Aydan 1996, 2002; Aydan and Hamada 1992; Aydan and Kumsar 1997; Aydan et al. 1996e, 2000a, 2000b, 2005a, 2005b, 2007a, 2007b, 2012a; Ergin et al. 1967; Emre et al. 1999; Eyidoğan et al. 1991; Gençoğlu et al. 1990; Ulusay and Aydan 2005; Ulusay et al. 2001, 2002, 2003a, 2003b). The data for other countries are those compiled by Wells and Coppersmith (1994), Matsuda (1981) and Sato (1989). The data on source properties of earthquakes are gathered from the well-known seismological institutes such as USGS, Harvard, ERI of Tokyo University, Swiss Seismological Institute. The number of the data set varies depending upon the studied empirical relations. For example, the number of data for the relation between M_w and M_s is 206. The following items are chosen as the characteristics of earthquakes:

1. Magnitude (moment and surface wave magnitudes, *Mw, Ms*).
2. Length of earthquake fault (*L*): *L* denotes the length of source fault or that estimated by the ground surface trace observed in the field or aftershock distribution if the surface rupture is hindered by thick sedimentary deposits (e.g., 1992 Erzincan, 1998 Adana-Ceyhan earthquake).
3. Depth of earthquake hypocenter (*D*).
4. Rupture area (*S*): *S* denotes the ruptured area of the earthquake fault inferred from aftershock distribution or the multiplication of surface rupture length produced by the earthquake by its hypocenter depth with the assumption of a rectangular source area.
5. Net slip of the earthquake fault (*Umax*): *Umax* denotes the maximum slip along the slip direction.
6. Maximum ground acceleration and velocity (*Amax, Vmax*) (hypocenter distance is mostly in the range of 15–25 km).
7. Rupture mode–striation orientation.
8. Ratio of vertical maximum acceleration to the horizontal maximum acceleration (RVAHA).

It should be noted that the minimum value of M_w is assumed to be 0 in all-empirical relations presented hereafter, although some events may have negative values and such events have no physical significance in Earthquake Science and Engineering.

7.1.5.1 Relations between Surface Wave Magnitude and Moment Magnitude

Aydan (Aydan 1997; Aydan et al. 1996e) originally selected surface wave magnitude M_s in developing his empirical relations for Turkish earthquakes since many data based on surface wave magnitude Ms were available and the magnitude of earthquakes have not exceeded 8 so far. It is pointed out that surface wave magnitude M_s becomes unreliable if it exceeds the value of 8 (i.e., Fowler 1990). Furthermore, it is becoming more popular to use the moment magnitude M_w in place of the surface wave magnitude M_s as many seismological institutes have recently released moment magnitude data rather than surface wave magnitude data. Nevertheless, the moment magnitude data determined by various institutes for the same earthquake are not always the same. Furthermore, the moment magnitude data must be assigned to previous earthquakes before the development of moment magnitude concept. Kanamori (1983) suggested that the surface wave and moment magnitudes of earthquakes can be taken as equal to each other within the range of 5–7.5. Aydan (Aydan 1997; Aydan et al. 1996e) proposed the following relation between surface and moment magnitudes of earthquakes for Turkish earthquakes.

$$M_w = 1.044M_s \text{ or } M_s = 0.958M_w \tag{7.20}$$

However, Ulusay et al. (2004) recently suggested the following formula for their data set on Turkish earthquakes:

$$M_w = 0.6798M_s + 2.0402 \tag{7.21}$$

Figure 7.8 compares the data set of Turkish earthquakes compiled by the author, including all recent data on Turkish earthquakes and worldwide data, which are fitted to the following empirical relation:

$$M_w = 1.2M_s e^{-0.028M_s} \tag{7.22}$$

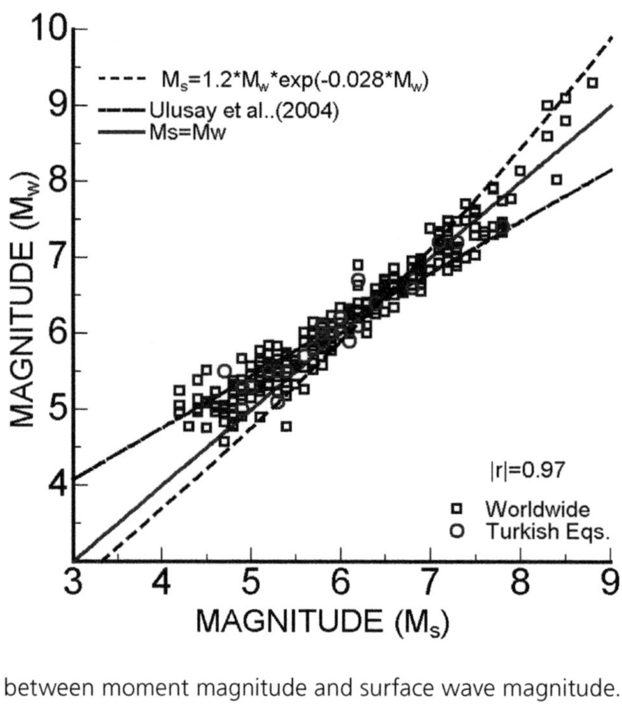

FIGURE 7.8 Relation between moment magnitude and surface wave magnitude.

As noted from the figure, all data generally support the suggestion of Kanamori (1983) for relating surface wave magnitude to moment magnitude for the magnitude range of 48. Therefore, it can be safe to adopt the previous empirical relations proposed by Aydan (1997, 2001, 2012a) based on surface wave magnitude for the seismic characteristics of Turkish earthquakes together with the replacement of surface wave magnitude with moment magnitude. However, the constants of functions should be recalculated if the independent variable is chosen as moment magnitude.

7.1.5.2 Relations between MMI Intensity and Moment Magnitude

The magnitudes of historical earthquakes are mainly inferred from the Modified Mercalli Seismic (MMI) intensity. Aydan et al. (1996e) and Aydan (1997) proposed the following empirical relation between MMI intensity (I_o) and surface wave magnitude M_s (see also Aydan 2001):

$$I_o = 1.317 M_s \tag{7.23}$$

In this study, the following relation between moment magnitude and MMI intensity is utilized with the consideration of recent large earthquakes in Turkey and worldwide:

$$I_o = 1.32 M_w \tag{7.24}$$

Yarar et al. (1980) and Gürpınar et al. (1979) also studied the relation between MKS intensity and earthquake magnitude for Turkish earthquakes. Their proposed relations were essentially similar to those given by Eqs. (7.23) and (7.24), except the constant has a minus sign. However, the magnitude in their formula is local magnitude. Furthermore, the coefficient for magnitude is about 1.6 times the one given in Eqs. (7.23) and (7.24). Kudo (1983) also studied the relation between MKS intensity and Intensity Scale of Japan Meteorological Agency. As the maximum values of intensity were different from each other, it can be said that roughly two intensity values of MKS would be designated, with one intensity value from the Intensity Scale of Japan Meteorological Agency. However, the intensity scales of 4, 5 and 6 of the Japan Meteorological Agency have been recently revised and subdivided as weak and strong intensity levels.

7.1.5.3 Relation between Moment Magnitude and Rupture Length, Area and Net Slip of Fault

It is natural to expect that the sense of faulting of earthquakes greatly influences their seismic characteristics. The faulting sense is considered in empirical relations between moment magnitude and rupture length (*L*), rupture area (*A*) and net slip (*Umax*) of fault given as an extension of the previous works of Aydan et al. (1996e) and Aydan (1997, 2007, 2012a) and as tested with the recent mega earthquake data (Aydan 2015c):

$$L, S \text{ or } U_{\max} = A \cdot M_w e^{M_w / B} \tag{7.25}$$

The functional form of the empirical relations is the same, while constants A and B differ depending upon the faulting sense, which are given in Table 7.2. Kudo (1983) and Ambraseys (1988) proposed similar empirical relations between rupture length and earthquake magnitude for Turkish earthquakes. Most earthquakes were due to strike-slip faulting. Although their relations are good fits to the data set used, they have limited applicability for the range of data set used in his study. Figure 7.9 shows the plot of data for the several parameters previously listed, together with empirical functions given by Eq. (7.25), Aydan (1997) and Well and Coppersmith (1994). The horizontal axis of the plots is the moment magnitude of earthquakes. As seen from the figure, the data are somewhat scattered. Nevertheless, the function, together with constants for each seismic parameter, is the best fit to observational data. The standard deviations of fitted equations to observational data were obtained and are shown in Figure 7.9. Furthermore, the relation proposed by Aydan (1997) without considering faulting sense is still valid, as noted from Figure 7.9, provided that the conversion of magnitudes is taken into account.

Another important source parameter is the duration of fault rupture. The duration is longer if the rupture propagation is unilaterally. However, it is shortened if bilateral rupture propagation takes place. Figure 7.10 shows a compilation of data on Turkish earthquakes and worldwide, including the most recent events such as the 2004 and 2005 Sumatra earthquakes and the 2005 Pakistan earthquake. The functional form of the empirical equation has the form of Eq. (7.25), and it is shown in Figure 7.10. together with empirical relations proposed by Dobry et al. (1978). The empirical relation shown in the figure holds for the data of Turkish earthquakes as well as worldwide data.

TABLE 7.2 Values of Constants for Eq. (7.25) for Each Fault Parameter

	RUPTURE LENGTH L (km)		*RUPTURE AREA S (km²)*		*MAXIMUM DISPLACEMENT U_{MAX} (cm)*	
	A	*B*	*A*	*B*	*A*	*B*
Normal faulting	0.0014525	1.21	0.003	1.5	0.0003	1.6
Strike-slip faulting	0.0014525	1.25	0.001	1.7	0.00035	1.6
Thrust faulting	0.0014525	1.19	0.0032	1.5	0.0014	1.4

7.1.6 Inference of Faulting Mechanism and Earthquakes

7.1.6.1 Inference from Striations of Earthquake Faults

The striations and internal structure of these faults are just evidence of what type of stress state caused them, and they may also indicate what type of earthquake they produced. The methodology for the inference of the possible stress state and focal plane solutions of earthquakes from the faults require data on dip, dip direction and striation orientation (Aydan 2000a; Aydan and Kim 2002; Aydan et al. 2002). Figure 7.11 shows an illustration of how striation angle is measured. Figure 7.12 shows the focal mechanism of earthquakes estimated from the fault striations obtained for the 1891 Kiso-beya earthquake in Japan and 1999 Kocaeli earthquake in Turkey.

7.1.6.2 Inference from Wave Propagation Characteristics

The focal mechanism of the earthquakes may also be inferred from the P-waves and S-waves induced by earthquakes. Figure 7.13 illustrates the concept of obtaining the focal mechanism of a vertical strike-slip fault and associated wave responses at the observation points around the focus of the earthquake. Figure 7.14(b) shows the focal mechanism solution for the motion of fault plane shown in Figure 7.14(a), and the seismograms for this solution would look as shown in Figure 7.14. Figure 7.15 shows the distribution of push-up and pull-down distributions to infer the faulting mechanism of the 1943 Fukui earthquake as an example.

The inclinations of fault planes have different orientations, and data-processing requires the depth, dip direction and takeoff angle of waves. For this purpose, a focal sphere fixed to the hypocenter of the earthquake, and the character of recorded waves and the location of the seismometer are considered.

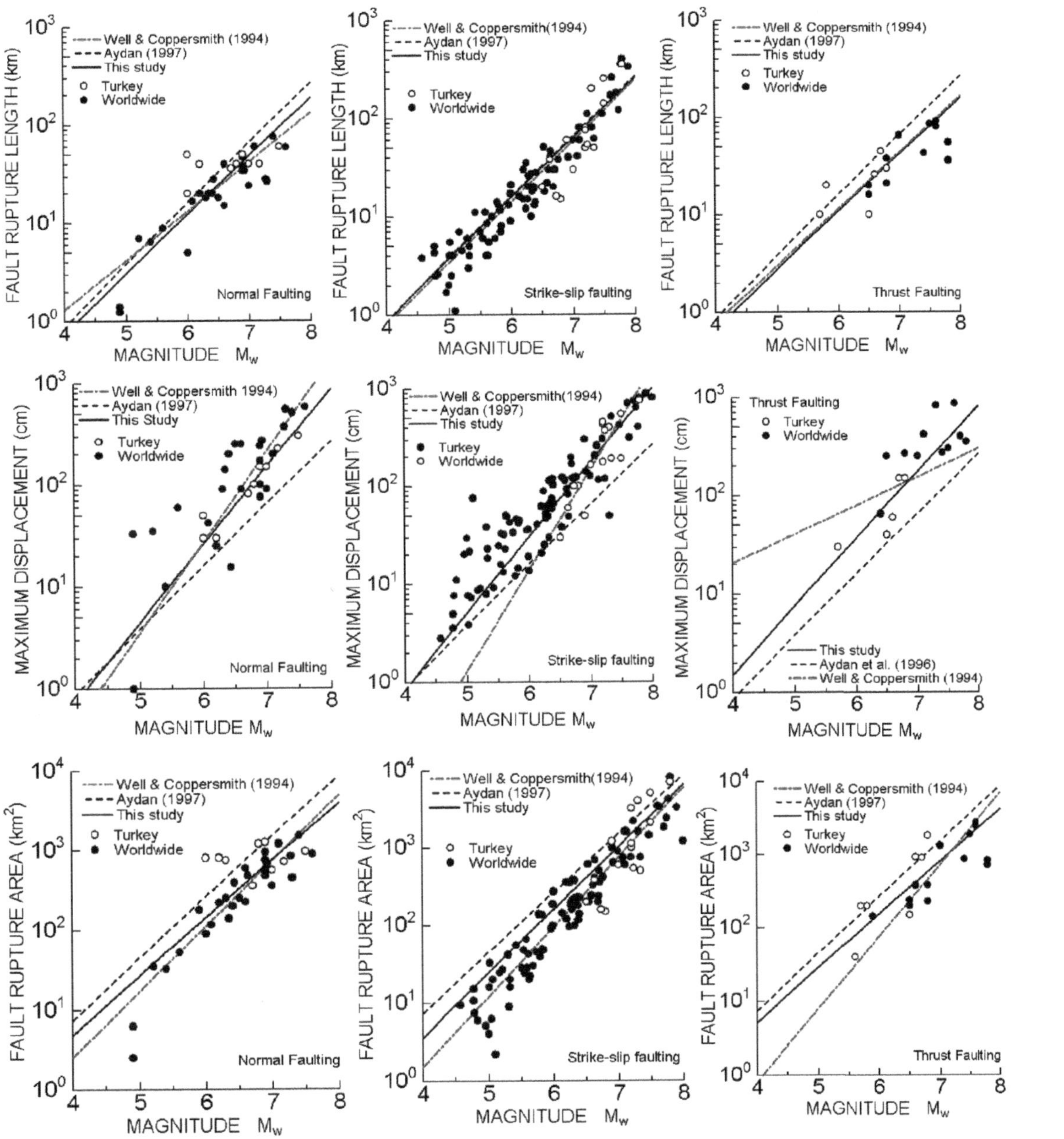

FIGURE 7.9 Comparison of various relations with observed data.

Figure 7.16 illustrates such an example for the sign and polarities of the P-wave first-motion seismic rays, while Figure 7.17 shows the focal mechanism of the 1984 Nagano-Seibu earthquake. Many seismological institutes provide the takeoff angle and the sense of motion data, which are utilized in obtaining focal plane solutions. However, it should be noted that a minimization procedure for observed data is necessary to determine focal plane solutions.

The focal plane solutions used in geoscience for inferring that the faulting mechanism of earthquakes is derived by assuming that a pure-shear condition holds. As a result of this assumption, one

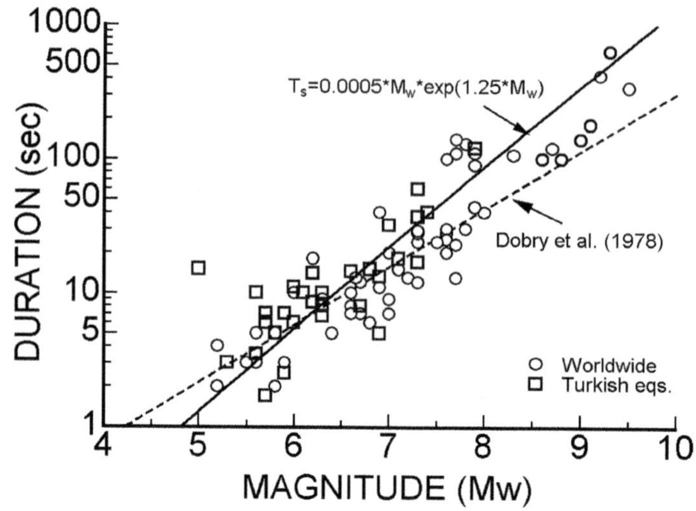

FIGURE 7.10 Comparison of various relations with observed data.

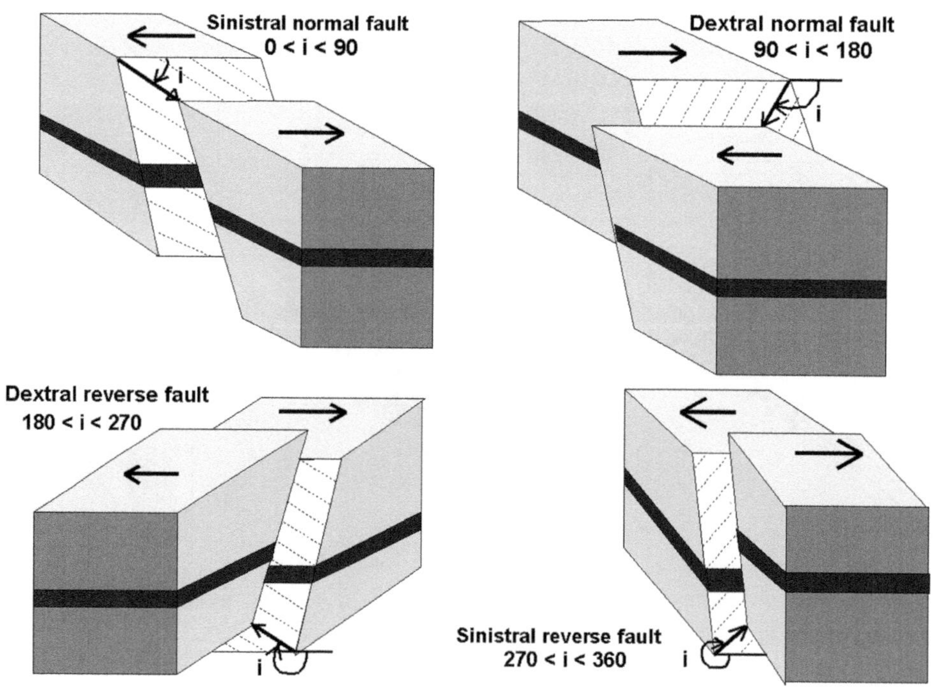

FIGURE 7.11 Definition of striation in structural geology.

of the principal stresses is always compressive while the other one is tensile in focal plane solutions. This condition may also imply that the friction angle of the fault is assumed to be nil. Therefore, the principal stresses are inclined at an angle of 45 degrees with respect to the normal of slip direction. This condition is used to determine P-axis and T-axis in focal plane solutions. Each focal plane solution involves the fault plane on which the sliding takes place and the auxiliary plane (Figure 7.18). The

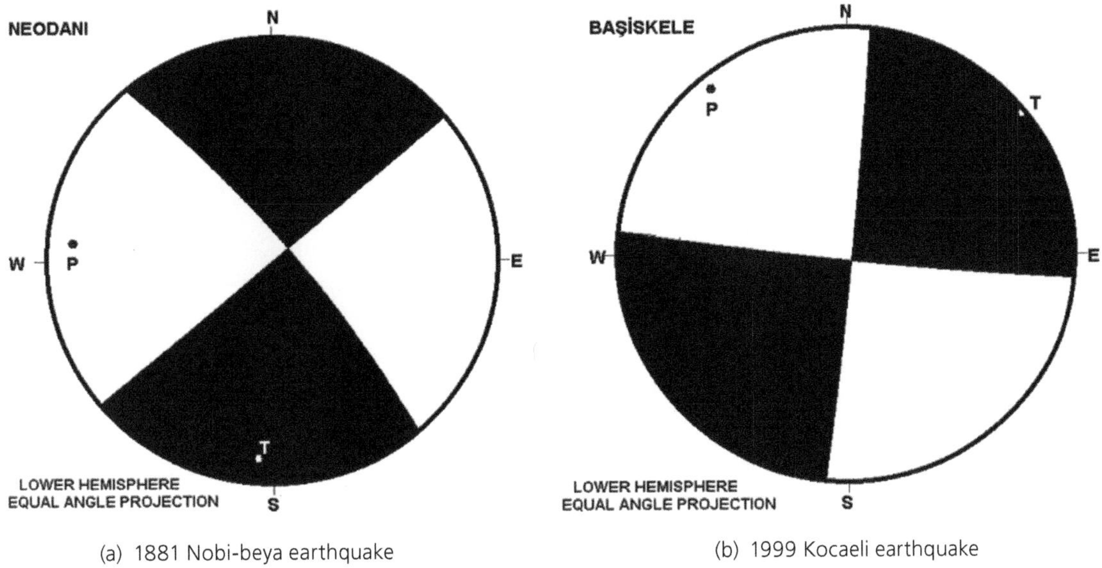

FIGURE 7.12 Inferred faulting mechanism for some earthquakes from fault striations.

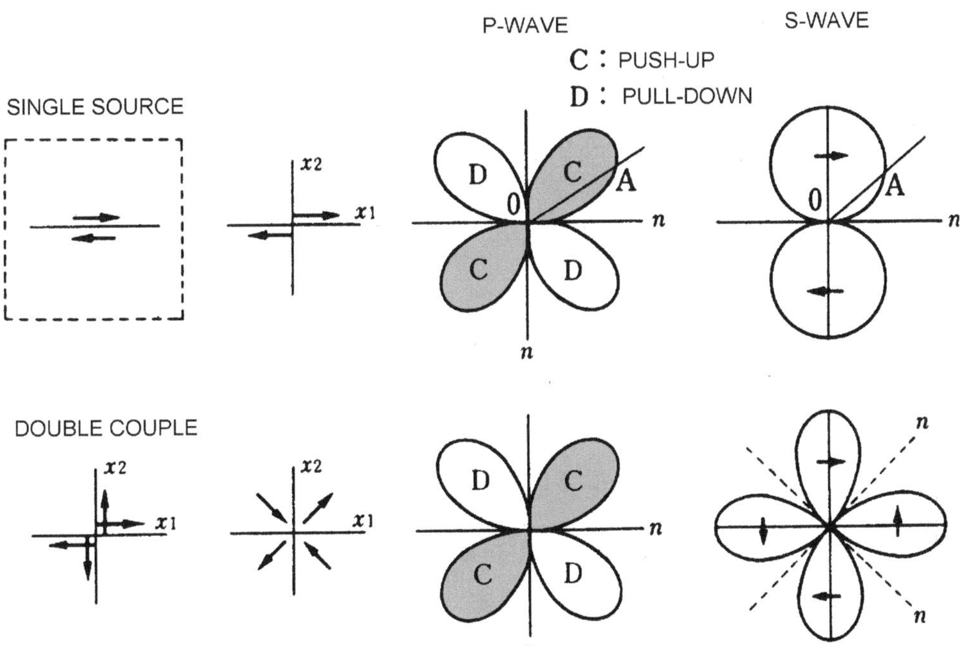

FIGURE 7.13 Illustration of fundamental concept to obtain focal mechanism solutions from seismic waves.

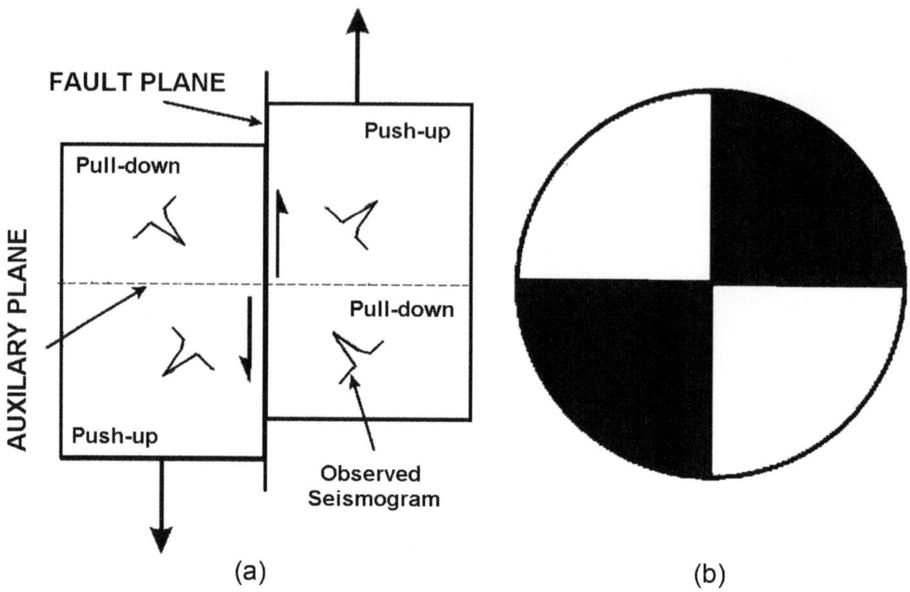

FIGURE 7.14 (a) Fault motion and (b) inferred focal mechanism.

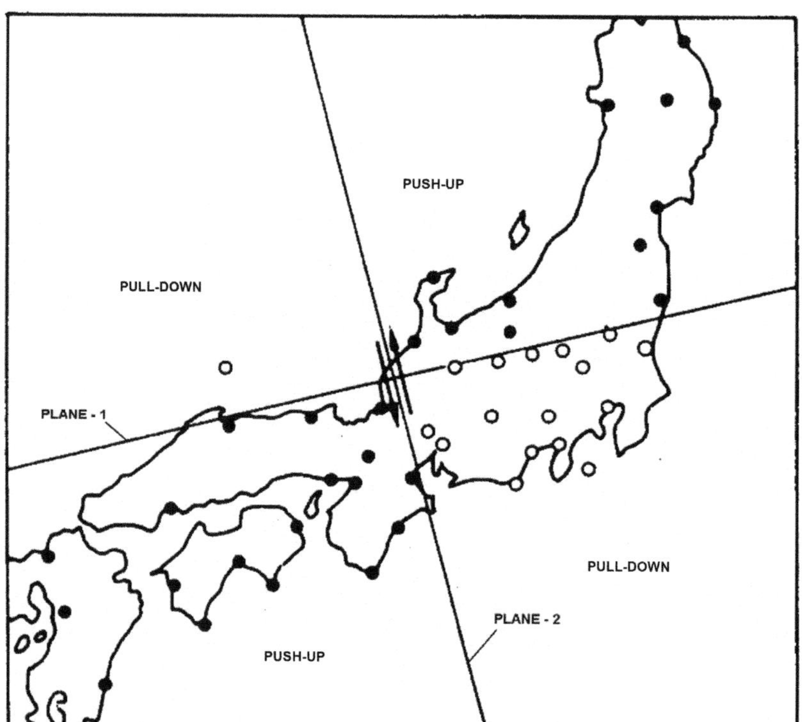

FIGURE 7.15 Focal mechanism of 1948 Fukui earthquake from pull-down and push-up distributions.

normal of the auxiliary plane corresponds to the slip vector, and it is orthogonal to the neutral plane on which the P-axis and T-axis exist.

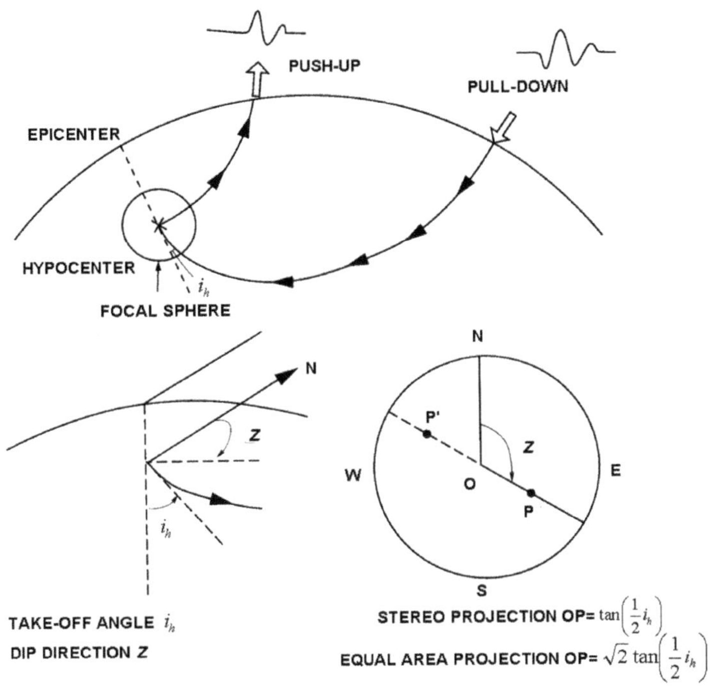

FIGURE 7.16 Sign and polarities of P-wave first-motion seismic rays (modified from Ando et al. 1996).

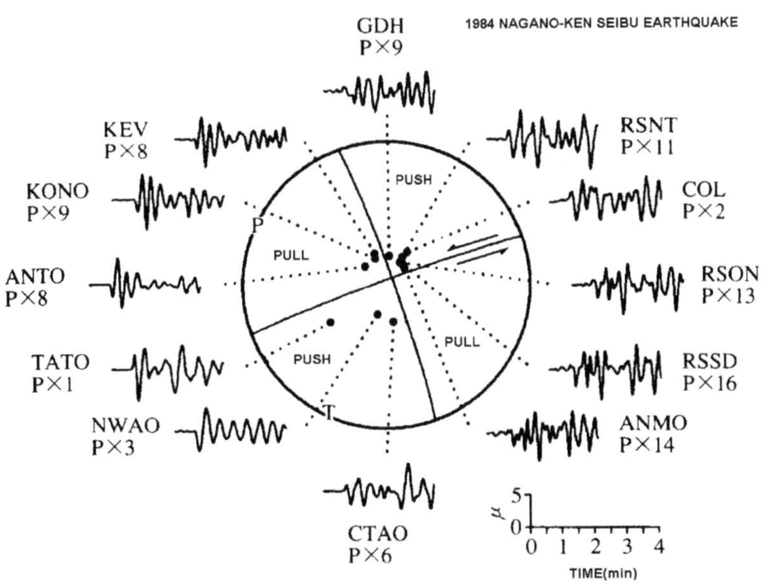

FIGURE 7.17 Focal mechanism solution for the 1984 Nagano-Seibu earthquake determined from P-wave first-motion seismic rays (modified from Ando et al. 1996).

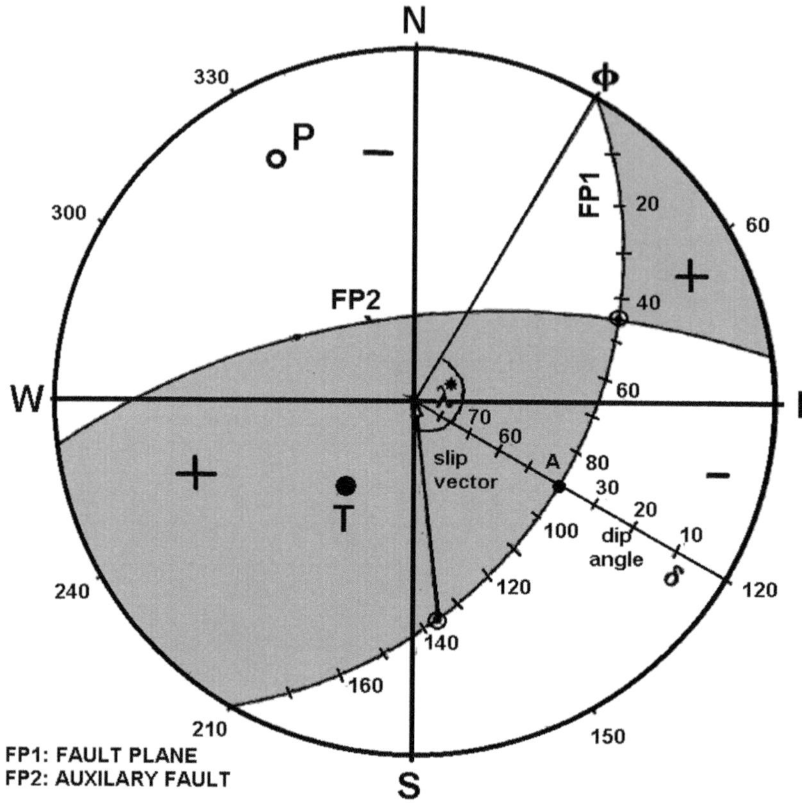

FIGURE 7.18 Illustration of fault plane, auxiliary fault, slip-vector on a lower hemisphere stereo net.

7.2 MOONQUAKES

The seismometers shown in Figure 3.11 recorded moonquakes, which were initially unexpected (Figure 7.19). The records from the moonquakes revealed the inner structure of the Moon like that of Earth. It was found that the earthquakes were categorized as deep earthquakes resulting from the tidal force interaction between Earth and the Moon, shallow earthquakes due to tectonic forces and thermal stress variations, and other earthquakes due the meteorite and lunar module impacts. The magnitude of the shallow earthquakes could be up to 5.5, while the magnitude of the deep earthquakes was much smaller and generally less than 2. The shallow earthquakes may pose a hazard to a lunar habitat, and they may last for more than 10 minutes. The deep earthquakes occur at depths of 700 to 1000 km, and they occur periodically at 14 and 206 days, implying the causes of the events are terrestrial and related to solar tidal forces. It was found that the vibrations caused by impacts can last for more than 1 hour while deeper earthquakes last only 10–20 minutes. The reason for such a long seismic vibration is considered to be the existence of a thick elastic crust (Figure 7.20). If the viscosity of the crust is low, the vibrations last longer for surface impacts.

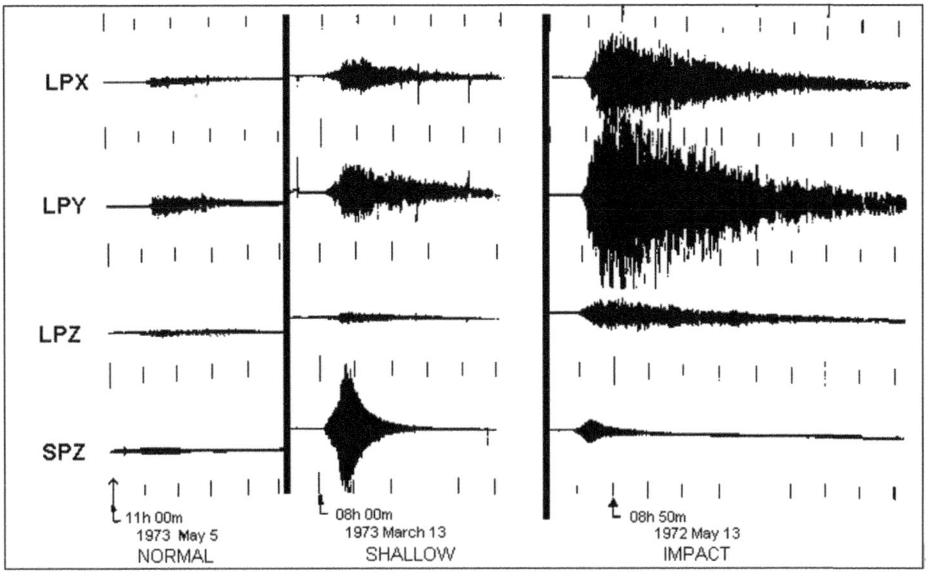

FIGURE 7.19 Records of moonquakes and lunar impacts (time marks 10 minutes) (from Nakamura et al. 1974).

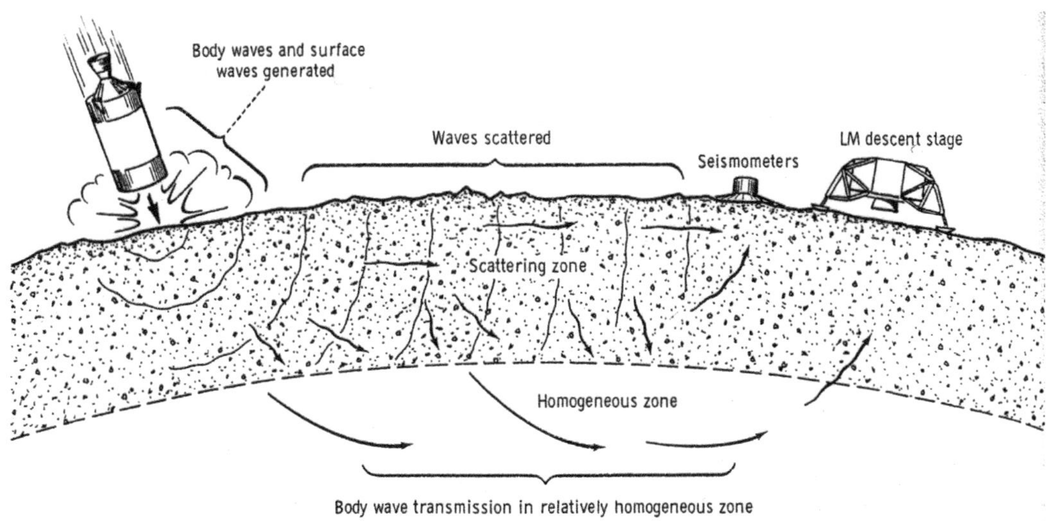

FIGURE 7.20 Illustration of the response of the Moon crust to impacts (Irvine 2002).

7.3 MARSQUAKES

As there were evidences of moonquakes from the Apollo program of NASA (1997), there was no doubt that Mars quakes should also occur (Anderson et al. 1977). Aydan (2017c, 2020a) stated the possible causes of Mars quakes.

The first seismic observation station, named InSight Lander, was installed on Mars very recently in 2018, when the InSight Lander touched down on Mars on November 27. Until the record of a seismic event on April 6, 2019, the occurrence of Mars quakes was a speculative issue. The first seismic event on April 6, 2019 confirmed the existence of quakes on Mars (e.g., Ceylan et al. 2023). The first vibration record obtained from a marsquake by the InSight Lander was quite longer than those observed on Earth and shorter than those on the Moon, as seen in Figure 7.21. The seismic monitoring would improve the understanding of the seismicity and the interior structure of Mars and provide essential data on the necessity of the quakeproof design of structures to be built in Mars or similar planetary objects. However, more similar instruments and instrumental data on Mars are necessary for decisive conclusions.

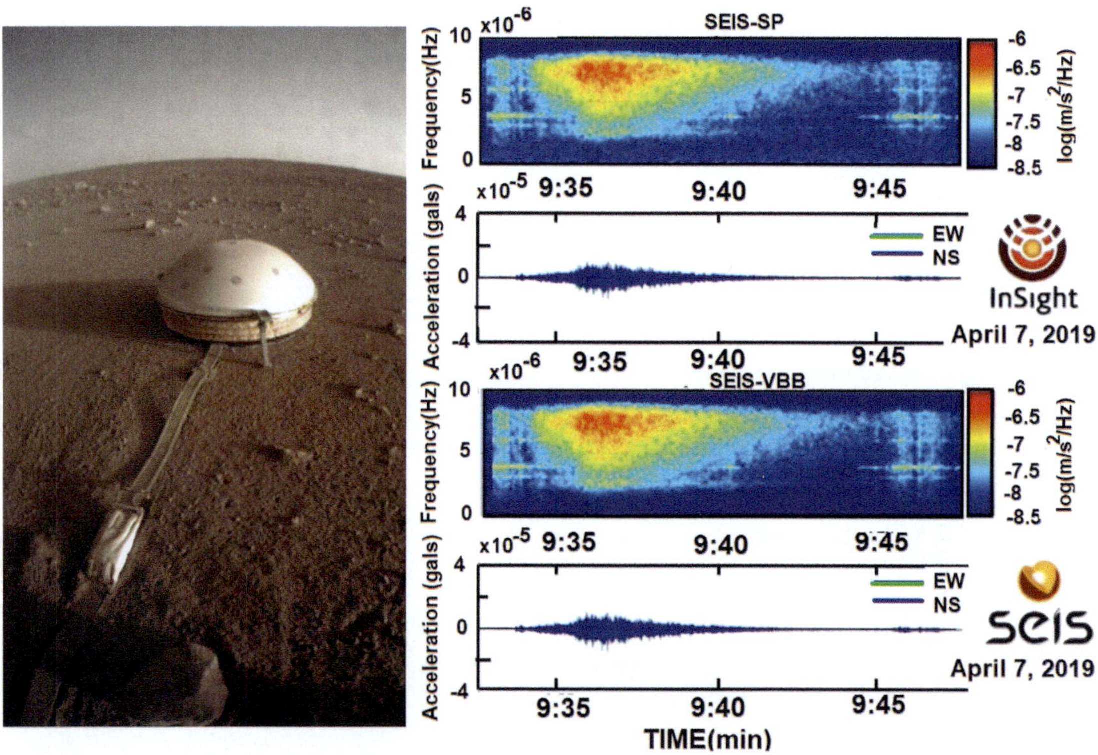

FIGURE 7.21 Seismic record by the seismometer of the InSight Mars module of a marsquake occurring on April 24, 2019 (arranged from NASA).

The InSight Lander recorded the largest marsquake with a magnitude of 4.7 on May 4, 2022. Figures 7.22 and 7.23 show the seismogram and its spectrogram recorded by the InSight Lander. These seismic signals were due to the impact of two meteoroids with an estimated diameter of 150 m, creating a crater with a diameter of at least 300 m. The InSight Lander terminated its recording task after the last communication on December 15, 2022, due to the power loss resulting from the dusting of its solar panels. The analyses of seismic records from the InSight Lander revealed that the core of Mars was smaller and denser.

Although it is extremely difficult to determine the focal mechanism of earthquakes with a single seismic station, there have been attempts to determine the focal mechanism of marsquakes. Brinkman et al. (2021) published the first attempts of focal plane solutions using the seismograms recorded by the InSight Lander, and they report that the focal mechanisms imply either normal faulting or strike-slip faulting mechanisms. However, they suggested the dominant faulting mechanism is normal faulting in the Cerberus Fossae region, which is known as a volcanic region with an extensional deformation mechanism. Nevertheless, thrust-type faulting also occurs. Figure 7.24 shows the locations of recent marsquakes reported by Ceylan et al. (2023).

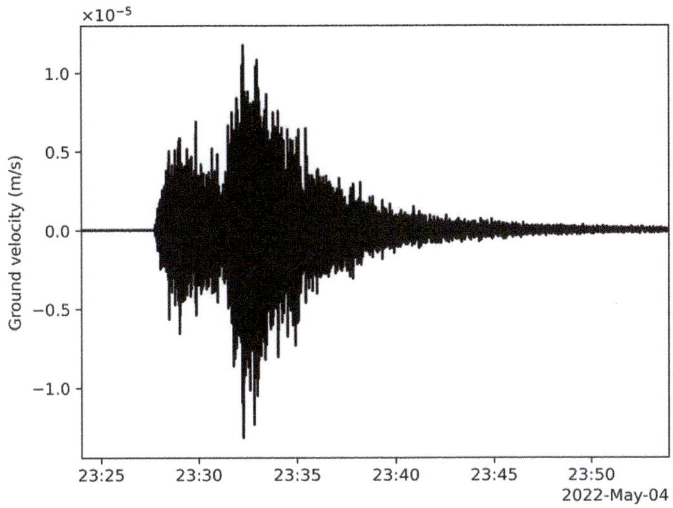

FIGURE 7.22 Seismogram of the largest quake ever detected on Mars on May 4, 2022 (from NASA/JPL-Caltech).

FIGURE 7.23 Spectrogram of the largest quake ever detected on Mars on May 4, 2022 (from: NASA/JPL-Caltech/ETH Zurich).

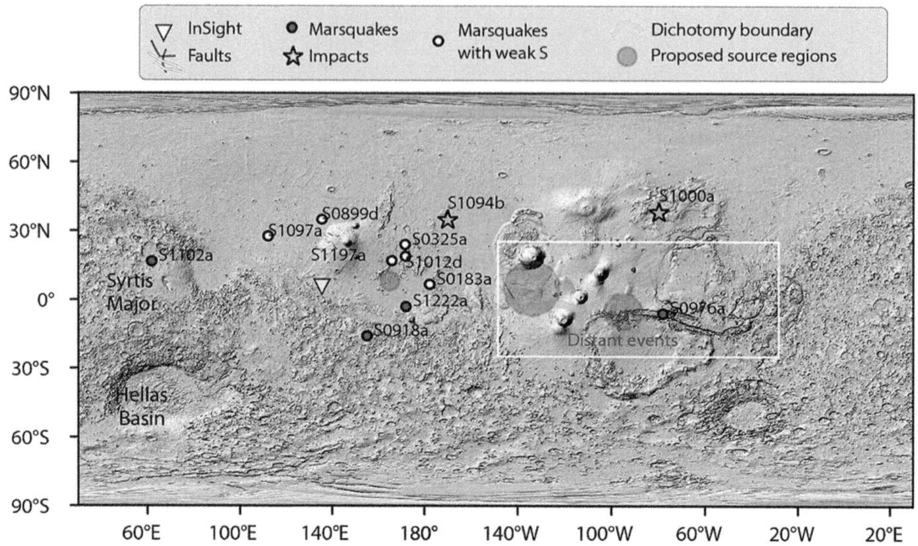

FIGURE 7.24 Locations of some recent marsquakes (from Ceylan et al. 2023).

7.4 VENUSQUAKES

There is no doubt that quakes do occur on Venus in view of volcanic activities and large-scale fracture zones seen in Figures 4.13 and 4.14. The possibility of installing seismometers on the surface of Venus is extremely difficult due to its thick clouds and greenhouse effect; surface temperatures of Venus approach 462°C, causing water vaporization. A new faulting event due to a venusquake in the same area was recognized by Messenger spacecraft flyovers between 1990 and 1994. The Soviets deployed instrumented balloons with pressure sensors from the two Vega lander craft immediately after entry into the atmosphere. The primary scientific objective of the Vega balloon probes was to obtain information about the large- and small-scale motions, structure and cloud properties of the Venus atmosphere at the float altitude (Sagdeev et al. 1986). Figure 7.25 shows an image of the ballon used in Vega projects by Soviets.

Recently NASA has been exploring utilizing the same technique to monitor earthquakes in California (Brissaud et al. 2021). Compared to conventional seismographs, there are some technical issues with balloons equipped with air-pressure sensors for determining the location and magnitude of earthquakes and their focal mechanisms, but it is probably the only technique available so far. An air-pressure sensor (7890 km away) at the home of the author with a sampling interval of 60 minutes could detect the Tonga volcano eruption at 13:10 JST (4:10 GMT) on 2021 January 15 at 20:31 JST on the same day after 441 minutes, as shown in Figure 7.26. In other words, this example confirms the possible use of air-pressure monitoring for such a purpose.

NASA also reported a detection of quake-induced mass movements from the SAR analyses by Messenger spacecraft in November 1990 soon after the start of mapping of the topography of Venus (Malin 1992). Although it is not very practical, the continuous SAR-type technique could be another approach to identify seismic events on Venus.

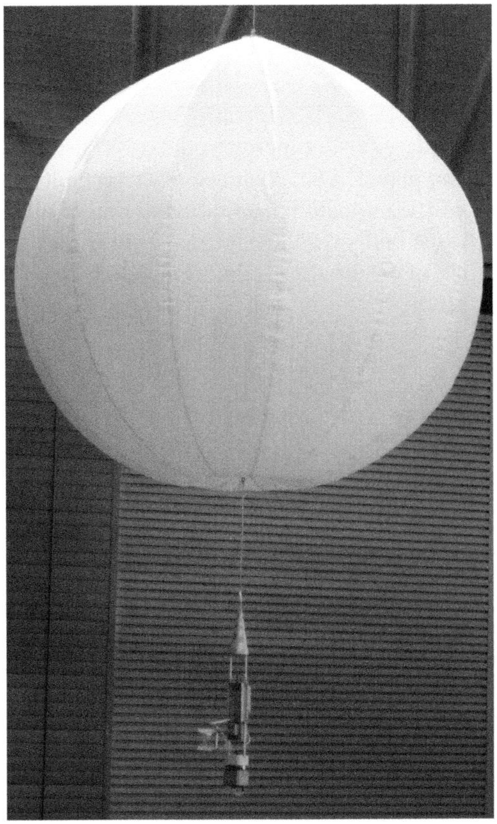

FIGURE 7.25 View of the balloon used in the Vega program.

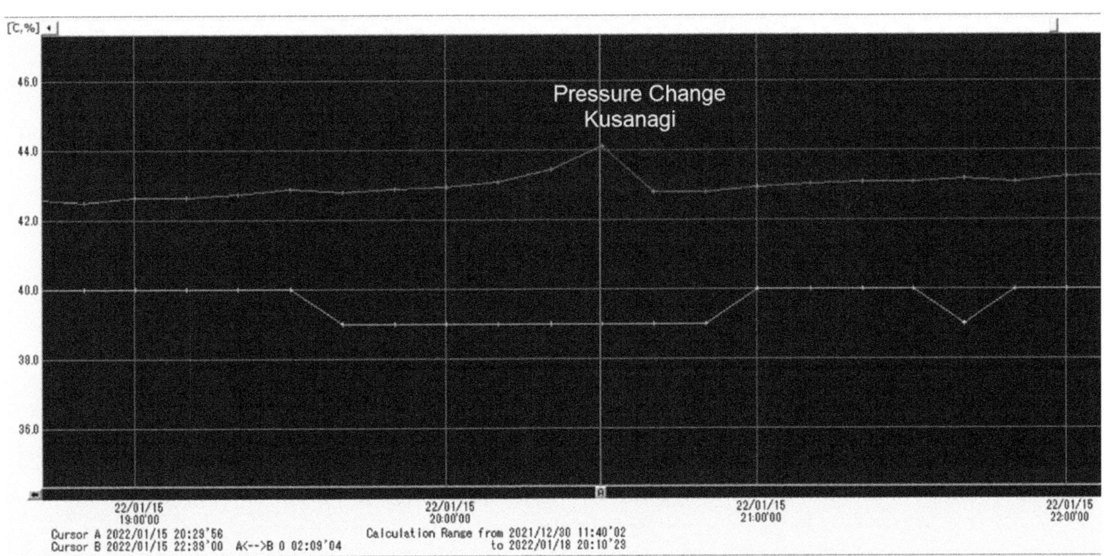

FIGURE 7.26 Pressure change at Kusanagi related to Tonga eruption on January 14, 2022.

7.5 MERCURYQUAKES

Mercury is likely to have seismic activities, or mercuryquakes. These quakes are thought to be primarily caused by the shrinkage of the planet. The existence of large fault scarps (up to 1000 km long), as seen in Figure 4.22, suggests significant tectonic activity and potentially strong "quakes" on Mercury. NASA detected smaller-scale thrust faults with a length ranging from 1 to 7 km. For such fault lengths, the estimated moment magnitude of earthquakes could range between 4.2 and 5.6 using the empirical relations proposed by Aydan (2012a).

Dynamics of Impacts and Crater Formation

8

The dynamic response of objects during free fall and impact to barriers may have some important implications in various fields, such as the collision of vehicles in transportation engineering, the collision of adjacent structures during earthquakes, standard penetration tests in soil mechanics, and impact craters due to meteorites, anchoring of ships or platforms in marine engineering and bullets and missiles destroying targets. Figure 8.1 illustrates several examples of the effects of impacts. This chapter is based on the relevant chapter in a textbook by Aydan (2017b), and it is concerned with the dynamics of impact phenomenon.

8.1 CRATER FORMATION BY METEORITES AND ITS ENVIRONMENTAL EFFECTS

Craters due to impacts on the surface of planets and their satellites are well-known, and there are some excellent books on this topic (i.e., Melosh 1989, 2011). Such craters are also found on the earth. It is well-known that nickel, copper, platinum, gold and palladium ores in Sudbury basin were caused by the impact of a gigantic meteorite. It is also well-known that the impact by a gigantic meteorite at Yucatan Peninsula (Mexico) caused the extinction of dinosaurs from the earth 66 million years ago. This crater is named the Chicxulub Crater with an estimated diameter of 180 km. The 1908 airburst of Tunguska event (M5) in Siberia caused heavy damage to wooden buildings and blew down 90% of the trees within a radius of about 15 km. A similar event took place again in Chelyabinsk region on February 13, 2013, causing damage to buildings and its surrounding due to the airburst (Figure 8.2). The estimated seismic magnitude is about 2.7 according to USGS. The most recent event of a meteorite fall occurred near Sarıçiçek village in Bingöl Province of Türkiye.

The meteorite impact on May 4, 2022, was caused by two meteorites, which caused a crater with a total diameter of 300 m. Views of the crater were captured by the Mars Orbiter, as shown in Figure 8.3. This event caused the largest quake in Mars recorded so far, and Figures 7.22 and 7.23 showed the seismogram and spectrogram recorded by the InSight Lander.

The cratering process has been well investigated by various researchers (i.e., Melosh 1989, 2011; Collings et al. 2005; Kenkmann et al. 2011), and Figure 8.4 shows an illustration of impact cratering. The impact mechanism by meteorite is divided into mainly three stages: (1) contact and compression, (2) excavation, (3) modification. Depending upon the size of the meteoroid, the craters are categorized as simple and complex, as illustrated in Figure 8.5.

The entry velocity of meteoroids ranges from 11 km/sec to 72 km/sec. However, the speed of the meteoroid rapidly decelerates as it penetrates into increasingly denser portions of the atmosphere. Due

DOI: 10.1201/9781003643852-8

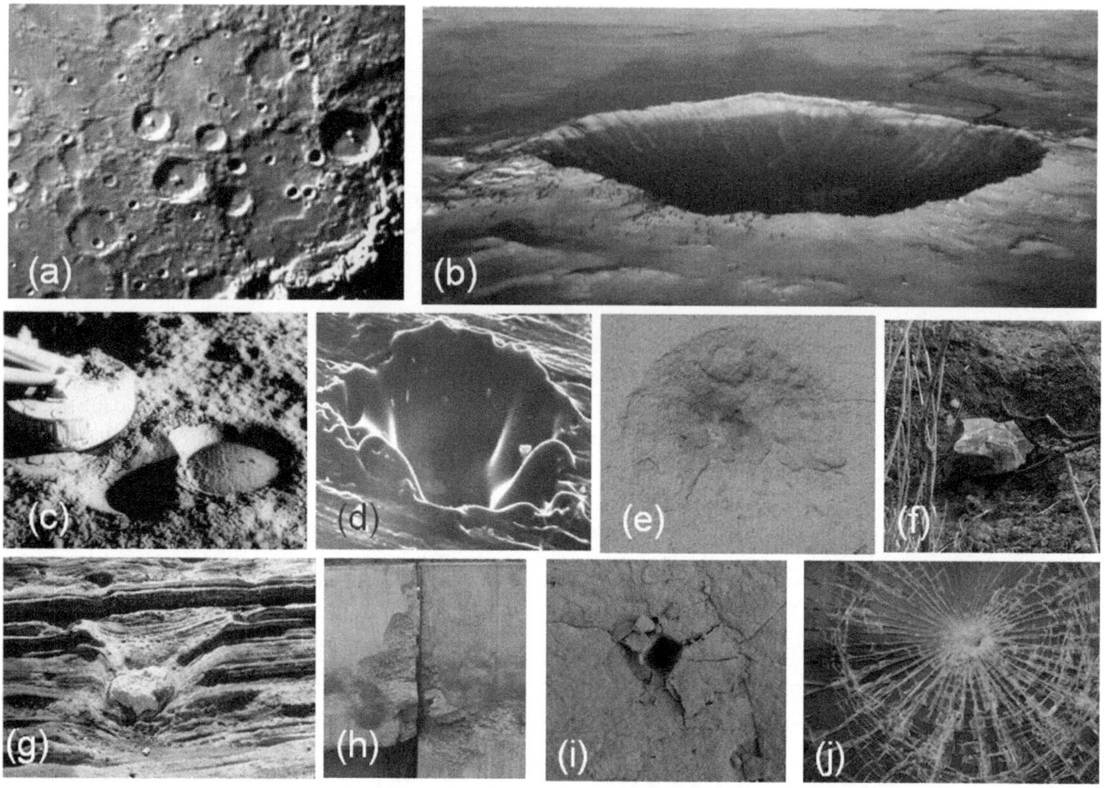

FIGURE 8.1 Examples of impacts: (a) Moon craters; (b) Arizona Meteor Crater; (c) Crater of Apollo landing pad; (d) small crater in glass of Apollo Module; (e) crater on the wall of Victoria and Albert Museum in London; (f) rock-fall crater caused by 2007 Noto Peninsula earthquake; (g) cratering in Volcanic sediments in Hawaii; (h) collision of bridge deck with abutment by 2008 Wenchuan earthquake; (i) crater induced by a geologic hammer in Shimajiri mudstone; (j) broken glass induced during 1999 Kocaeli earthquake. Figures (a)–(d) are from the images in the Smithsonian Air and Space Museum taken by the author during his visits, (e)–(j) are author's pictures.

to the atmospheric drag of the Earth, the weight of most meteorites ranges from a few kilograms up to about 7000 kg and its terminal velocity is about 90–180 m/s when it hits the ground. However, as the atmosphere of Mars is thin, these values would be quite different.

Kenkmann et al. (2011) conducted laboratory experiments with dry and wet sandstone blocks impacted by centimeter-sized steel spheres. They utilized a powerful two-stage light gas gun to achieve impact velocities of up the 5.4 km/s. They concluded that cratering efficiency, ejection velocities, and spall volume are enhanced if the pore space of the sandstone is filled with water. In addition, the crater morphologies differed substantially in both experiments, and the vaporization of water upon pressure release significantly contributed to the impact process. Strength and elastic modulus of sandstone used in experiments were 62.4±2.8 MPa and 14.8±1.4 GPa for the dry and 47.0±3.7 MPa and 12.1±1.0 GPa for a water-saturated sandstone, measured perpendicularly to the bedding planes.

Major terrestrial craters found on Earth are well-documented (e.g., Osinki 2006). Figure 8.6 shows the distribution of major craters on Earth. However, it should be noted that there may be some hidden large craters in seabeds or beneath the ice-covered north and south poles, which are unnoticed so far.

FIGURE 8.2 Some views of the meteorite fallen in the Chelyabinsk region on February 13, 2013, and its damage (from AP Photo; see also IRIS 2013).

FIGURE 8.3 Impact craters caused by meteorites on May 4, 2022, on Mars.

8.2 EFFECTS OF METEORITES

Effects of terrestrial impacts on the Earth may be categorized as airburst, seismic, dust-fall and tsunami. In this subsection, the state of art on these aspects is summarized. The readers are advised to consult the quoted references for further information.

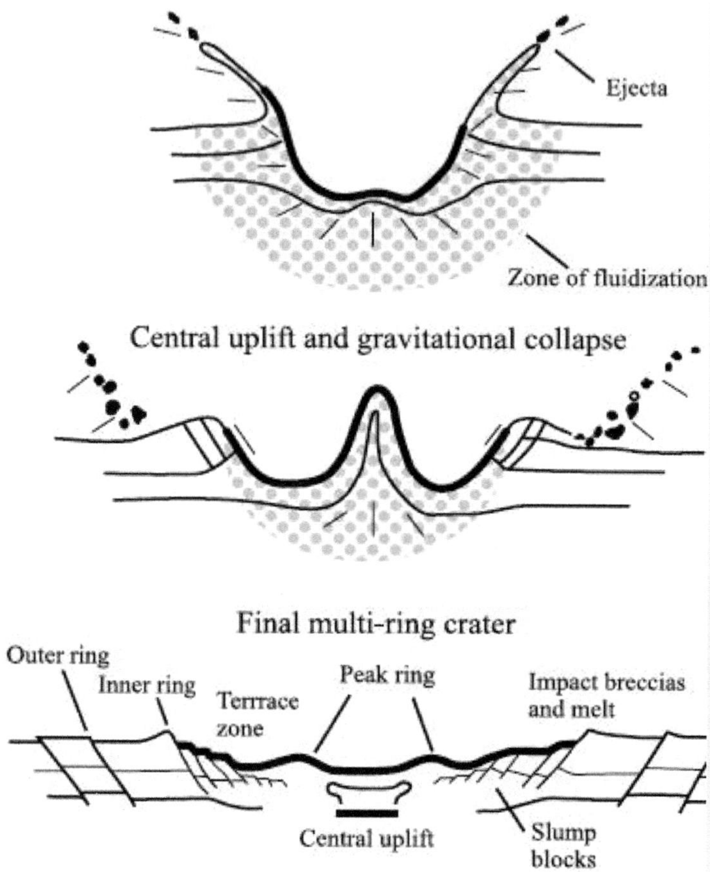

FIGURE 8.4 Illustration of cratering process (from Melosh 1989).

Collins et al. (2005) summarized some fundamental formula to estimate the geometry of craters, seismic effects, dust-fall induced by the impact of meteoroids. The kinetic energy (K) of an impactor with mass (m_i) and velocity (v_o) just before atmospheric entry can be written as:

$$K = \frac{1}{2} m_i v_o^2 \tag{8.1}$$

If the impactor is generally assumed to be a spherical body, its initial mass is given as:

$$m_i = \frac{\pi}{6} \rho_i D_o^3 \tag{8.2}$$

Density of the earth's atmosphere is assumed to be decaying outward from the given density at the surface of the earth, and it is given by the following equation:

$$\rho = \rho_o e^{-z/H} \tag{8.3}$$

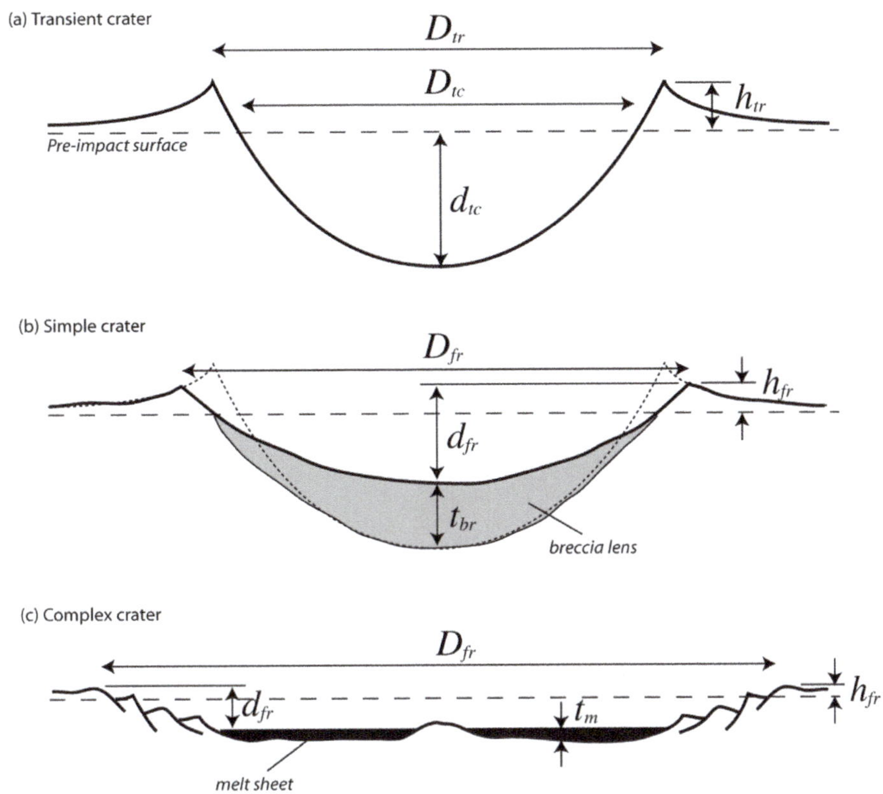

FIGURE 8.5 Illustration of cratering process (from Collins et al. 2005).

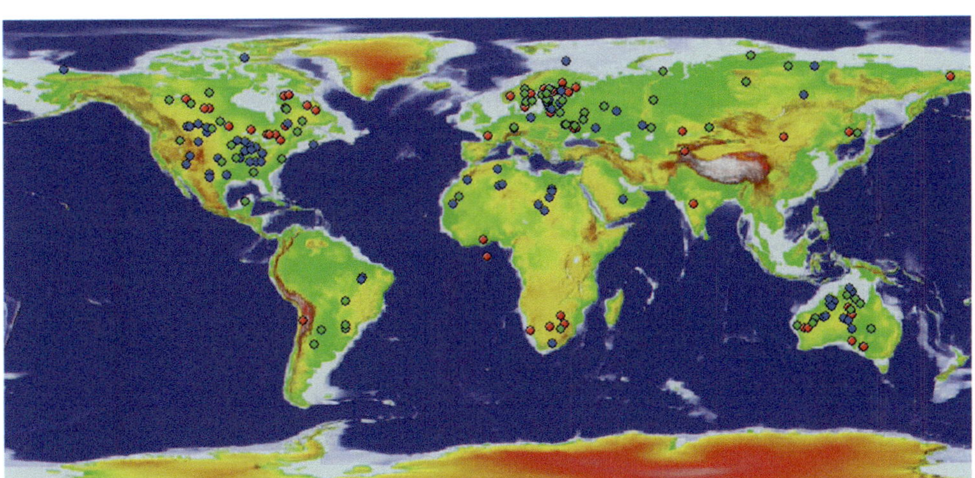

FIGURE 8.6 Distributions of terrestrial impact structures (from Osinki 2006).

The velocity of the impactor as a function of altitude is derived on the basis of exponential decaying of the density of Earth's atmosphere is given by:

$$v(z) = v_o \exp\left\{ -\frac{3\rho(z)C_D H}{4\rho_i D_o \sin\theta} \right\} \tag{8.4}$$

The estimation of the geometry of craters caused by meteorites is a very complex problem. The following relations are for the diameters of transient, simple and complex craters, its depth and volume of breccias:

Transient Diameter

$$D_{tc} = 1.161 L^{0.78} v_i^{0.44} g_E^{-0.22} \sqrt[3]{\frac{\rho_i}{\rho_t}} \sin\theta \tag{8.5}$$

Simple Crater Diameter

$$D_{fr} = 1.25 D_{tc} \tag{8.6}$$

Volume of Breccias

$$V_{br} = 0.032 D_{tc}^3 \tag{8.7}$$

Depth of Breccias

$$d_{fr} = 0.4 D_{fr}^{0.3} \tag{8.8}$$

Seismic effect is related through the empirical relation given by Gutenberg–Richter magnitude (M)-energy relation:

$$M = 0.67 \log_{10} E - 5.87 \tag{8.9}$$

where E is assumed to be equal to the kinetic energy of the impactor expressed in Joules. The peak overpressure decreases exponentially with distance from ground zero:

$$p = p_o e^{-\beta r_1} \tag{8.10}$$

where p_o and β are functions of burst altitude given by:

$$p_o = 3.14x10^{11} z_b^{-2.6} \tag{8.11}$$

where

$$\beta = 34.87 z_b^{-1.73} \tag{8.12}$$

$$r_1 = \frac{r}{E_{kT}^{1/3}} \tag{8.13}$$

8.3 CRATER FORMATION BY PROJECTILES IN ROCKS

Craters produced by missiles with oblique trajectories and kinetic energies between 2.1×1014 and 81×1014 ergs produce craters 2–10 m across. The craters and their ejecta are typically bilaterally symmetrical about the plane of the trajectory. Crater dimensions are strongly dependent on target material and kinetic energy of the missile. Craters in wet targets are significantly larger than those in dry porous targets (Moore 1971). The explosion velocity of missiles generally ranges between 71 m/s and 204 m/s.

Bullets or shrapnel have similar characteristics, and the size of craters depends upon the characteristics of target material as well as the size of bullets or shrapnel. Figure 8.7 shows several pictures of craters on the wall rocks of the Victoria and Albert Museum in London induced by the shrapnel of World War II bombs. The craters are generally 100–150 mm in diameter having a depth of 15–30 mm, and the wall rock is Portland stone, which is a Jurassic aged limestone with a uniaxial compressive strength of about 70–80 MPa. The bullet muzzle velocity is in the range of 1067 m/s to 1524 m/s, while the conventional artillery cannons range between 914 m/s to 1067 m/s.

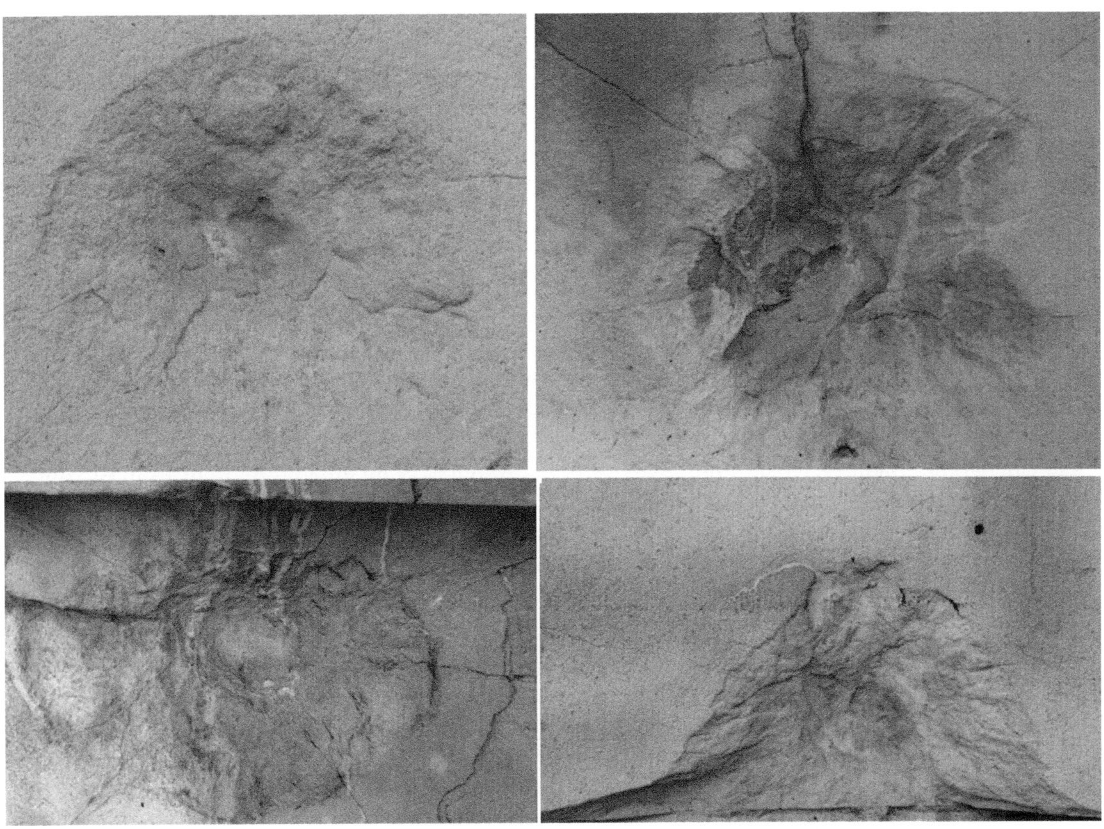

FIGURE 8.7 Craters on the wall of Victoria and Albert Museum induced by the bombshells.

8.4 MONITORING OF VIBRATIONS CAUSED BY METEORITES

There is almost no remarkable seismic record of meteorite impacts on Earth. The record reported by Roelofse and Saunders (2013) is probably the first seismic record of a meteorite. The record is shown in Figure 8.8, and it was measured by the Mussina seismic station of South African National Seismograph Network. The signal consists of two distinct parts: a signal originating at 20:55:37.59 (UT) with a duration of approximately 77 s (Figure 8.8), which will be referred to as S1, and a second signal (S2) commencing almost 138 s after the first at 20:58:11.08 (UT). The second signal's duration is approximately 193.14 s.

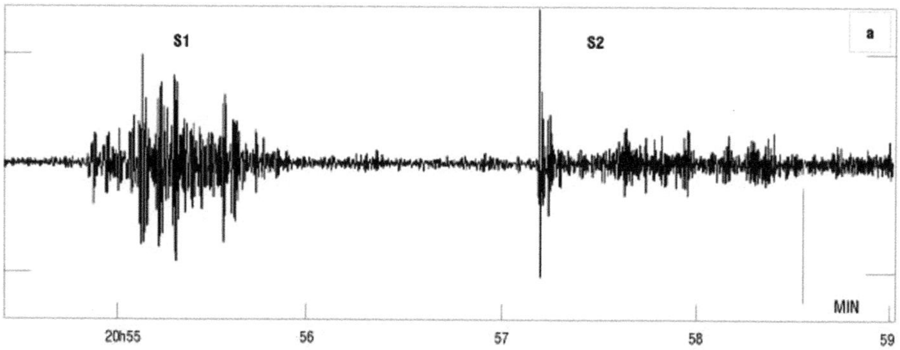

FIGURE 8.8 Seismic record at the Mussina seismograph of a meteorite (from Roelofse and Launders 2013).

As mentioned in the previous subsection, the recent meteorite impact in Chelyabinsk on February 13, 2013, induced seismic surface waves observed at five seismograph stations of the IRIS/USGS Global Seismographic Network (IRIS 2013) (Figure 8.9). The nearest station is denoted by ARU, which is about 160 km away from the impact location, in addition to the infrasound detectors installed at USArray stations extending from the midcontinent to the eastern seaboard. The USArray recordings from the meteor show complex arrivals lasting for almost an hour. The entry mass was 13000 tons with a diameter of about 20 m and a density of 3.3 g/cm^3. A 7 m hole was discovered in 70 cm thick ice on Lake Chebarkul. Small meteorite fragments were recovered over an area up to 50 m from the impact location. It is estimated that a 200–1000 kg meteorite would be required to create such a hole. A mass of ≥570 kg was recovered from the lakebed.

NASA had installed some seismometers to observe earthquakes and meteorite impacts on the Moon. Since 1970, four stations with seismic sensors have been placed on the Moon at Apollo 14, 15, 16, and 17 sites (i.e., Toksöz 1975; Nakamura 2003; Nakamura et al. 1975). When combined with the seismic station operating at the Apollo 12 site since November 1969, these have formed an excellent network. The network has detected a large number of moonquakes and meteoroid impacts. In addition to these, there have been nine artificial impacts (ascent stage of the lunar module [LM] and Saturn third-stage booster [S4B]) of known energies, impact times, and locations (Toksöz 1975). The records from meteoroid impacts and the lunar module (LM) of Apollo 12 are shown in Figure 8.10. Due to the inherent characteristics of the interior structure of the Moon (Latham et al. 1971; Toksöz et al. 1972, 1974), it takes a very long time for the dissipation of surface waves induced by surface impacts on the Moon.

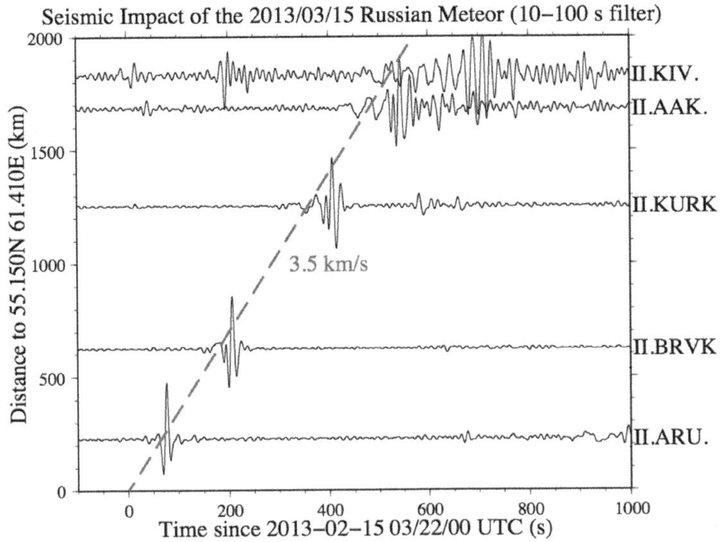

FIGURE 8.9 Surface wave records at several seismograph stations induced by the Chelyabinsk Meteorite (IRIS 2013).

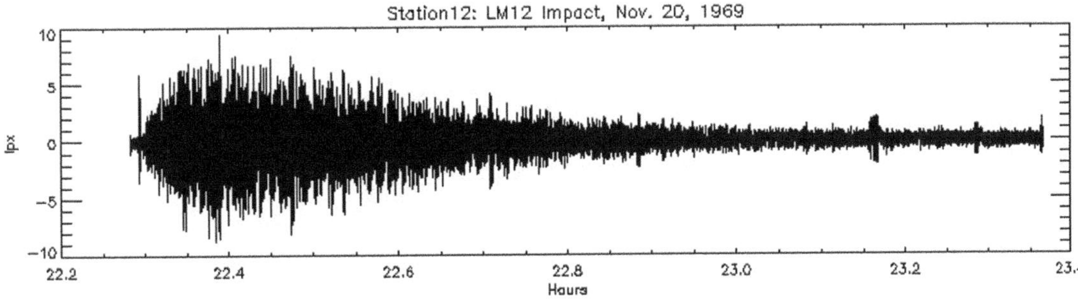

FIGURE 8.10 Seismic records at Station 12 on the Moon by Lunar Module 12 (Lindsay 2008).

8.5 FREE FALL (DROP) EXPERIMENTS

The dynamic response of objects during free fall and impact may have some important implications in earth science and geo-engineering. In this subsection, some free fall (drop) experiments under gravitational environment are presented (Aydan 2017b).

8.5.1 Objects Falling on Dry Sand Layer or a Mixture of $BaSO_4$, ZnO and Vaseline Oil

The experimental setup consists of the cylindrical and spherical objects with given dimensions and a container with dry sand or a mixture of $BaSO_4$, ZnO and Vaseline oil, as illustrated in Figure 8.11. The falling object and container are equipped with accelerometers that can measure the acceleration up to 12 G. The physical and strength characteristics of sand are given in Table 8.1, and its grain-size distributions are shown in Figure 8.12. During tests, the size of the objects and free fall height were varied, and accelerations of the impact object and container were measured. In addition, the embedment depth following the impact was also measured. The objects were concrete, glass, marble and iron. Figure 8.13 shows views of some of experiments.

TABLE 8.1 Physical and Mechanical Properties of Dry Sand

MEAN GRAIN SIZE (mm)	*DRY UNIT WEIGHT (kN/m³)*	*POROSITY (%)*	*FRICTION ANGLE (°)*
0.46–0.60	14.7	43.73	32–34

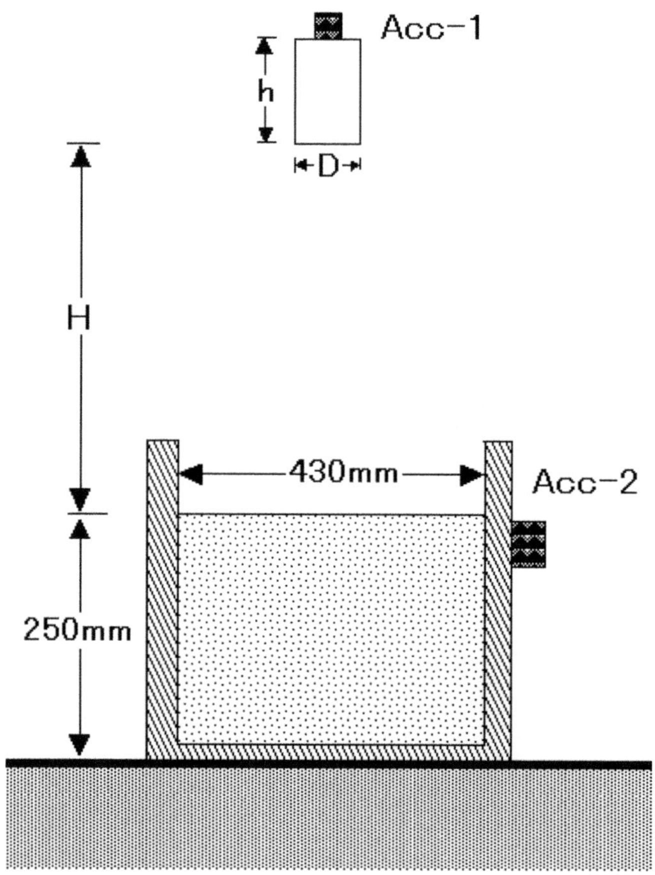

FIGURE 8.11 Experimental set-up for free fall and impact tests.

8.5.1.1 Experiments with Concrete Cylinders

Concrete cylinders had a diameter of 100 mm and height of 200 mm. Its weight was 3750 gf (unit weight: 23.87kN/m^3). It was dropped from a height of 1000 mm on a dry sand layer. Three tests were carried out. Figure 8.14 shows measured accelerations during one of the experiments. It is of great interest that the maximum acceleration of the concrete cylinders is almost same and is about 11.6–11.7 g.

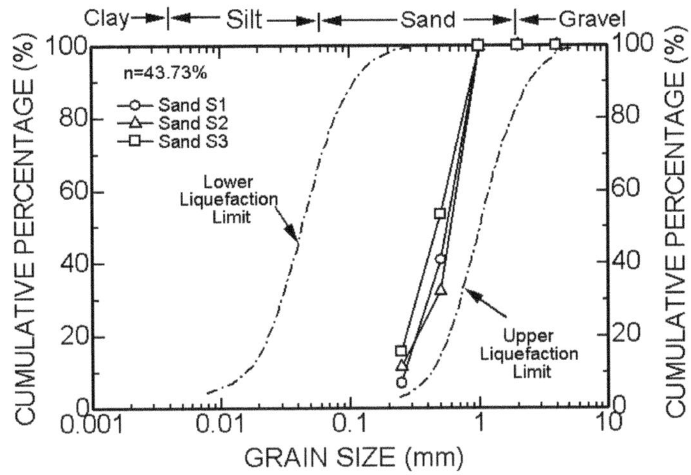

FIGURE 8.12 Grain size distributions of the sand.

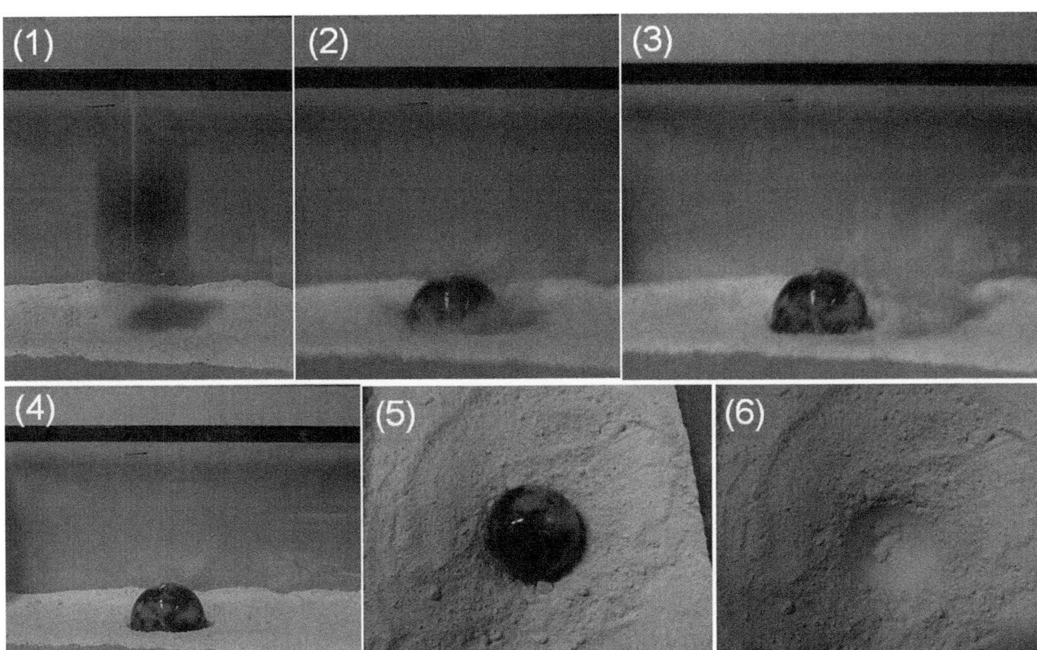

FIGURE 8.13 Views of a marble sphere during different stages of drop and its crater.

When the object is released, the vertical acceleration reaches to gravitational acceleration, and it remains to be the same during free fall as expected. The maximum ground acceleration occurs during impact, and the generated wave form is not symmetric. The ground acceleration at a distance of R/a = 4.3 is about 0.6–0.72 g, and its wave form is symmetric. From the experimental results, it seems that the following expression holds:

$$ma_{\max} = (m + M)a_g \tag{8.14}$$

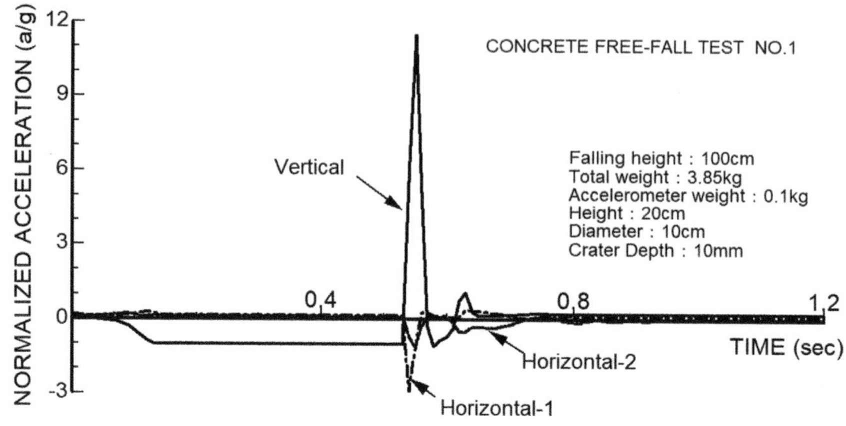

(a) Three components of concrete cylinder

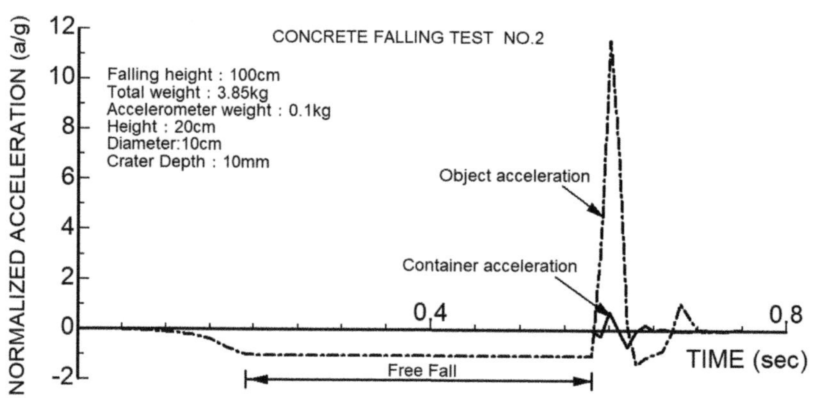

(b) Vertical accelerations of the object and the container

FIGURE 8.14 Acceleration response during free fall and impact (concrete cylinders).

where m and M are the masses of the object and ground permanently displaced, and a_{max} and a_g are the accelerations of the object and the ground at the time of impact. The cratering depth was 10 mm, and it was almost the same in all experiments.

8.5.1.2 Experiments with Iron Cylinders

Iron cylinders had a diameter of 25 mm and height of 100 mm. It was dropped from heights of 500, 1000 and 1500 mm on a dry sand layer. Three tests were carried out. Figures 8.15 and 8.16 show measured accelerations during the experiments. It is of great interest that the maximum acceleration of the iron cylinders varies, that it is more than 11 G, and that it seems the vertical accelerations are the same for three different drop heights, as seen in Figure 8.16. When the object is released, the vertical acceleration reaches gravitational acceleration, and it remains the same during free fall as expected. However, the cables of the accelerometer might impose some resistance to the magnitude of the maximum acceleration during free fall.

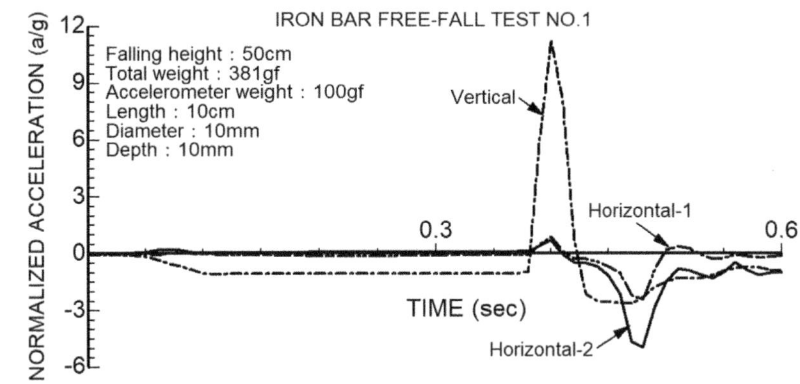

FIGURE 8.15 Acceleration responses during free fall and impact (iron bars).

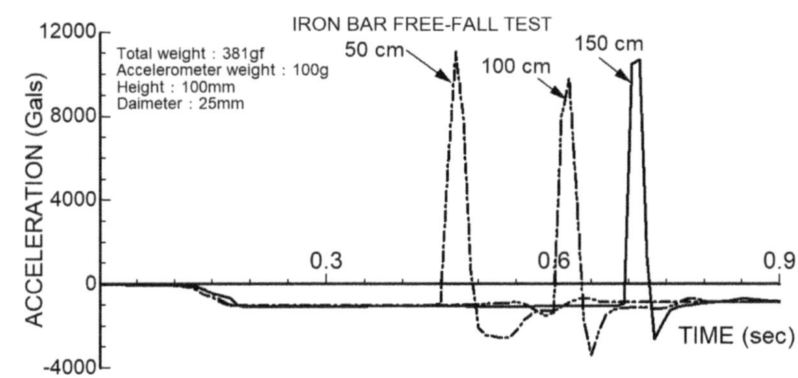

FIGURE 8.16 Vertical acceleration responses during free fall and impact (iron bars) (drop height is variable: 50, 100 and 150 cm).

8.5.1.3 Experiments with Marble Sphere and Glass Sphere

A series of experiments were carried out on a marble sphere with a 80 mm diameter and a glass sphere with a diameter of 66 mm. The spheres were dropped from a height of 900 mm on a loosely compacted layer of a mixture of $BaSO_4$, ZnO and Vaseline oil. Accelerations were measured using stand-alone type accelerometers developed by the author. Views of the marble sphere during different stages of free fall are shown in Figure 8.13. All stages of crater formation depicted by Melosh (1989) were clearly observed in the tests despite the color of base ground, which was whitish.

Figures 8.17 and 8.18 show the acceleration responses measured during free fall and impact. As noted from the figure, the vertical acceleration is equal to gravitational acceleration as expected but becomes more than 2.4 times the original acceleration. Furthermore, the horizontal accelerations are also remarkably high at 2.3–2.5 times the gravitational acceleration. The penetration depth ranged between 28 and 31 mm for spheres. The penetration depth in the loosely compacted layer of a mixture of $BaSO_4$, ZnO and Vaseline oil was remarkably higher than that in the dry sandy layer. As a result, the maximum accelerations observed during impact were quite lower in the loosely compacted layer of a mixture of $BaSO_4$, ZnO and Vaseline oil. This implies that the ground motions caused by the impactor in soft ground are less as compared with those on hard ground.

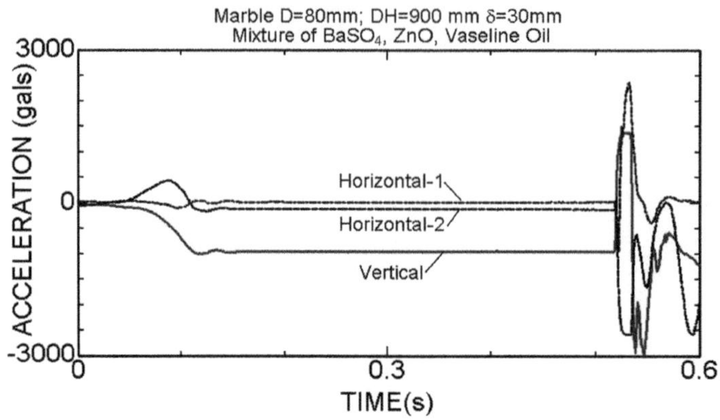

FIGURE 8.17 Acceleration responses during free fall and impact (marble sphere).

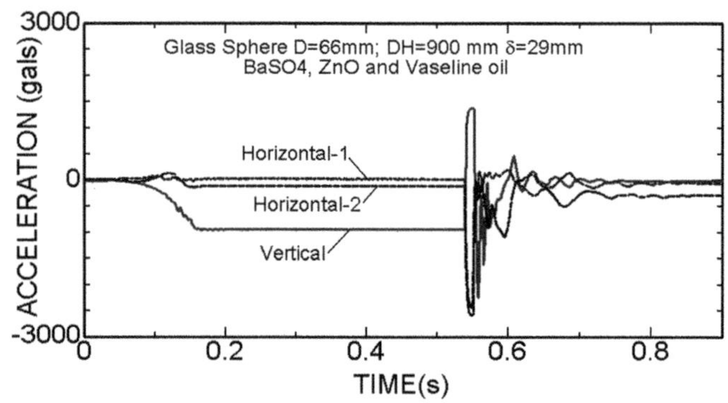

FIGURE 8.18 Acceleration responses during free fall and impact (glass sphere).

8.5.1.4 Crater Formation

The impact of objects causes cratering in the ground, as illustrated in Figure 8.19. The energy dissipated through the bearing failure of the ground beneath the object and elastic wave propagation. The energy (E) of the falling object under the gravitational field at the time of impact may be written in terms of the fall height (H) and the weight (W) of the object:

$$E = W \cdot H \tag{8.15}$$

The energy intensity (e) per unit area (A_c) of the object projected in the direction of movement may be defined as follows:

$$e = \frac{E}{A_c} \tag{8.16}$$

Figure 8.20 shows the relation between the energy intensity and embedment depth for cylindrical steel and tuff samples on dry sand. The falling direction was perpendicular to the longitudinal axis of the samples. It seems that penetration depth, normalized by the diameter of the object, is influenced by the material type of the objects.

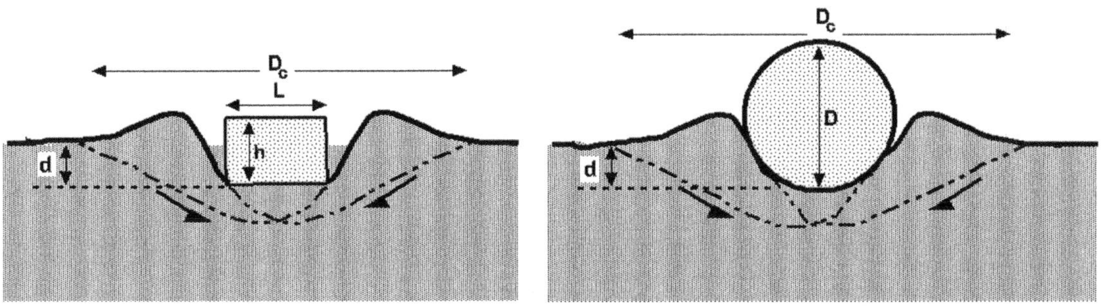

FIGURE 8.19 Types and parameters of craters created by the impact of falling objects.

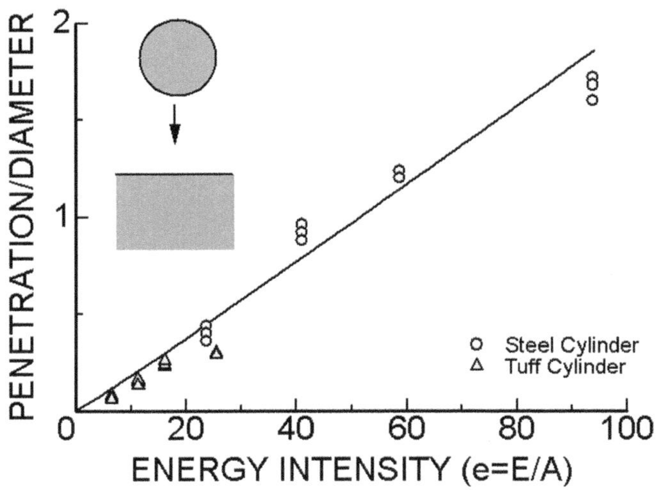

FIGURE 8.20 Relation between energy intensity and penetration depth.

8.5.2 Sandbags Falling on Hard Base

The author with his colleagues carried out some impact tests by dropping a 1000 kgf sandbag onto a hard base from a height of 1m and 1.8m in relation to the assessment of impact forces on a large underground substructure built for the protection of karstic caves by landing airplanes (Aydan and Tokashiki 2011). A total of ten accelerometers (six accelerometers were compact stand-alone type) were used, and one of accelerometers was attached to the falling object and four of the accelerometers were installed in the underground structure, which were 23 m from the drop location. Views of a drop test are shown in Figure 8.21. Five different layout patterns of drop tests were designed.

The maximum acceleration was about 23 g when the sand bag hits the hard base (Figure 8.22). However, the maximum acceleration quickly decreases with distance. Taking into account possible hemispherical wave propagation from the drop location, the measured maximum accelerations were fitted to the following empirical relation.

$$a_{\max} = 20000e^{0.8*W*h}\frac{1}{1+r^3} \tag{8.17}$$

FIGURE 8.21 Views of a drop test.

where r, W and h are radial distance (m), drop object weight (kgf) and drop height (m), respectively. The unit of acceleration is gals. Figure 8.23 compares estimations by Eq. (8.17) with measured maximum accelerations.

The measurements for all patterns of instrumentation including measurements in underground structure are plotted in Figure 8.23. Eq. (8.17) was modified:

$$a_{\max} = Ae^{0.8*W*h}\frac{1}{1+r^{b}} \tag{8.18}$$

where b and A are empirical constants. The value of b ranges between 2 and 3, while the value of A is 20000. When b is 2, it corresponds to cylindrical attenuation. On the other hand, if the value of b is 3, it corresponds to spherical attenuation. When the measurements shown in Figure 8.22 are taken into account, the best fit was given when the value of b is 2.5.

The drop height was increased from 1 m to 1.85 m, and the validity of the attenuation relation given by Eq. (8.18) was checked. Figure 8.25 compares the measurements with the estimations from Eq. (8.18). Despite some scattering of measurements, the empirical relation could be used for estimating maximum accelerations with distance. However, it should be noted that the value of A would probably depend upon the stiffness of the base.

8.5.3 Drop or Backhoe Impact Test at a Bridge Foundation Site

Some impact tests by a backhoe bucket were carried out at a bridge foundation site consisting of weak sandy-gravelly limestone formation with the purpose of investigating the effect of piles of the

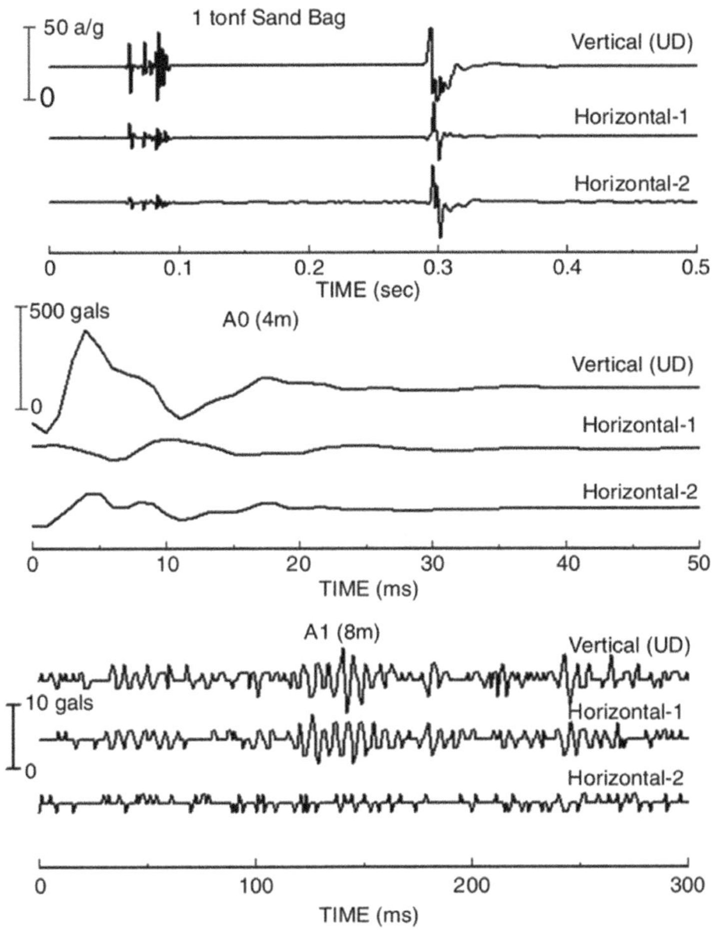

FIGURE 8.22 Acceleration records measured in a Pattern-0 drop test.

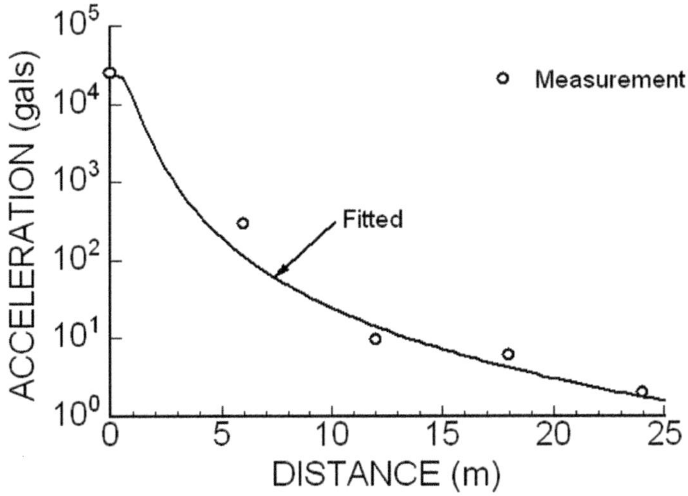

FIGURE 8.23 Comparison of estimations with measurements for Pattern-0.

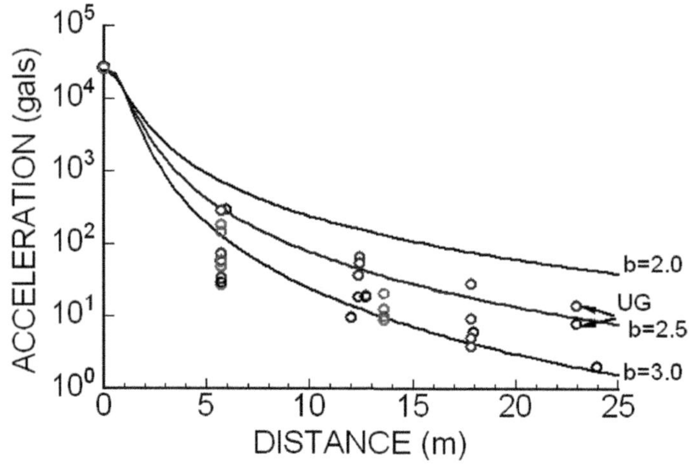

FIGURE 8.24 Comparison of estimations with measurements.

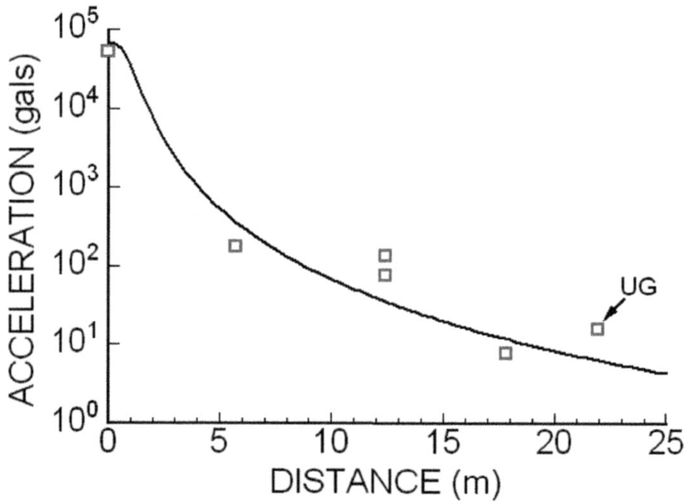

FIGURE 8.25 Comparison of estimations with measurements (drop height is 1.85 m, and the value of *b* is 2.5).

abutment of a bridge. Five accelerometers spaced at a distance of roughly 5 m with a sampling interval of 50 μs. Figure 8.26 shows the locations of the underground cave, impact and accelerometers on a cross section of the construction site. The nearest accelerometer to the impact location is about 5 m, while the farthest accelerometer is about 25 m. Figure 8.27 shows the acceleration records of the devices located at a distance of 5, 11 and 15 m, respectively. As anticipated from previous investigations, the amplitude of accelerations decreases with distance. Although the acceleration of the impact induced by the backhoe bucket was not measured, the overall tendency is similar to that estimated from Eq. (8.18). Furthermore, the P-wave velocity was estimated to be ranging between 670 m/s and 2000 m/s (Figure 8.28). The velocity of the top layer is much higher than that of the bottom.

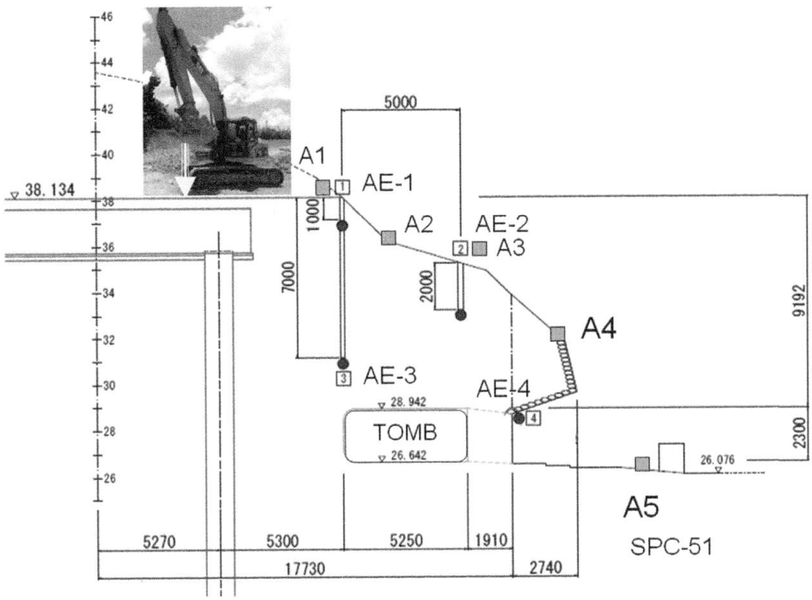

FIGURE 8.26 Location of instruments, impact, and underground structure.

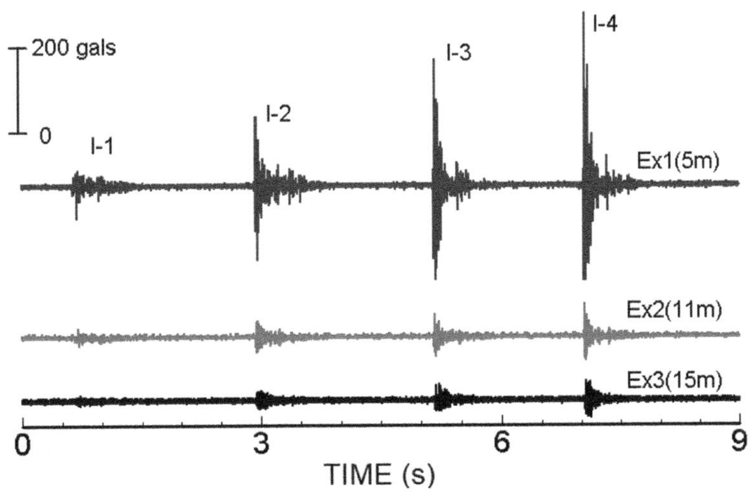

FIGURE 8.27 Recorded accelerations during impact monitoring at a limestone construction site (from Aydan 2017b).

8.6 PENETRATION OF CYLINDRICAL PROJECTILE INTO ROCKS IN DROP WEIGHT TEST

In this subsection the penetration response of cylindrical objects into rock under different impact velocities using a drop-weight device developed by the author is explained. First, paraffin was selected as

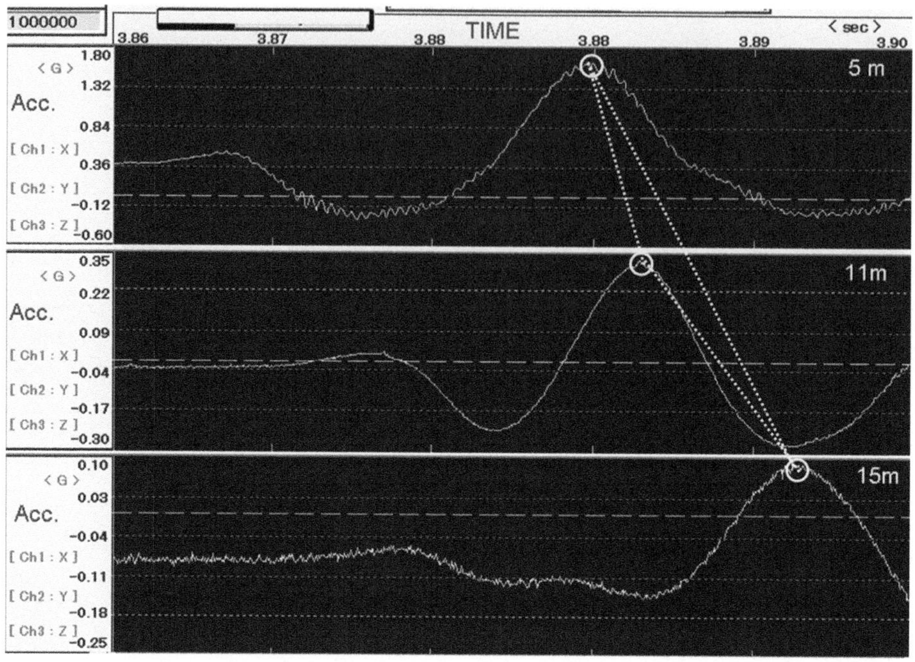

FIGURE 8.28 Close-up of acceleration records induced by the third impact in Figure 8.27.

a frictionless elastic-perfectly plastic material, while Oya tuff and Ryukyu limestone were chosen as brittle rocks. The drop height was varied in order to investigate the effect of impact velocity as well as the momentum imposed on samples. During experiments, the load and accelerations induced by shock loads are simultaneously measured, and the fracturing state was recorded through high-speed video cameras.

8.6.1 Shock Device and Theoretical Considerations

An experimental apparatus was developed to investigate the behavior of rocks under shock waves, as shown in Figure 8.29. The device equipped with load cell, noncontact laser transducers, accelerometer up to 500 g. The noncontact-type laser displacement transducers measure the displacement of the loading platen. The load cell is capable of measuring much higher dynamic loads. The displacement of the loading platen is allowed to move downward to 20 mm in order to prevent the total destruction of samples upon failure. The cylindrical weight (5, 10 and 15 kgf) can be dropped from different heights up to 500 mm with intervals of 50 mm. The device is fundamentally categorized as the drop-weight apparatus, and it is possible to evaluate the mechanical behavior and characteristics of rocks subjected to shock waves during pre-failure as well as post-failure stages. Some additional monitoring is done using infrared camera and high-speed video camera.

The maximum nominal velocity at the time of impact on samples can be computed from the following formula:

$$V_{max} = \sqrt{2gH_d} \tag{8.19}$$

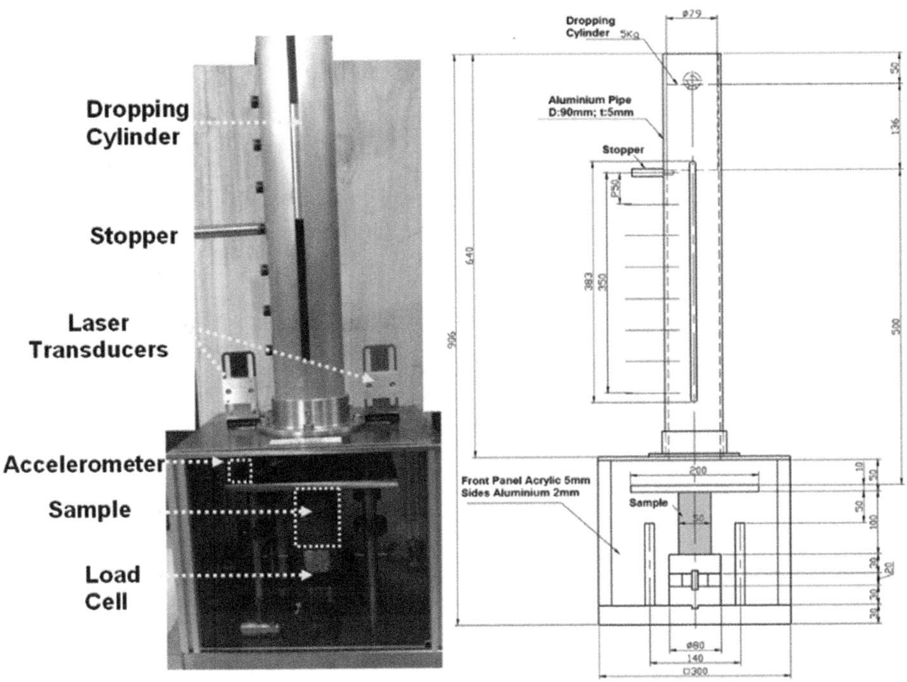

FIGURE 8.29 View (left) and schematic drawing (right) of the testing device.

where g is gravitational acceleration, and H_d is drop height. In this study, we also define maximum nominal strain rate by dividing the maximum nominal impact velocity by the sample height for the uniaxial testing condition:

$$\dot{\varepsilon}_{\max} = \frac{V_{\max}}{L} \tag{8.20}$$

Maximum acceleration is obtained from the acceleration response during the experiment. It is expected that the maximum acceleration increases with the increase of the strength of rock samples. On the basis of the theoretical definition of momentum, the theoretical relation for momentum (P) applied to a sample may be written in the following form:

$$P = W\sqrt{\frac{2H_d}{g}} \text{ or } P = W\frac{V_{\max}}{g} \tag{8.21}$$

where W is the weight given in kgf or N. The unit utilized in processing experimental results is $N \cdot s$.

8.6.2 Materials and Projectile

Paraffin, Oya tuff and Ryukyu limestone were used in experiments, and the impact device was hardened steel.

8.6.2.1 Paraffin

Paraffin was the pioneering material for researchers to study the fracturing and formulating elasto-plasticity and elasto-visco-plasticity. In a nonlinear regime, its behavior is perfectly plastic, and its friction angle is known to be nil. This material was chosen to investigate the effects of various parameters such as drop height and weight in order to evaluate the effect of momentum imposed on impact objects. Furthermore, the material is commercially available at very low cost. The diameter of the samples was 50 mm, and their height was 50 mm. Table 8.2 gives the physico-mechanical properties of the paraffin samples tested.

TABLE 8.2 Physico-mechanical Properties of Paraffin

γ (kN/m³)	V_p (km/s)	V_s (km/s)	UCS (MPa)
0.89–0.9	1.86–2.19	0.79–1.05	1.71

8.6.2.2 Rock Samples

Ryukyu limestone is widely distributed in the Ryukyu Archipelago. Depending upon the formation process, it is fundamentally divided into coral, sandy and conglomeratic limestone. Samples of sandy (locally known as Awaishi) Ryukyu limestone, Oya tuff, were prepared. The nominal size of samples was 50×50×50 mm. Table 8.3 summarizes mechanical properties of rock samples.

8.6.2.3 Projectile

The impact object has the shape of cylinder with a diameter of 10 mm and height of 35 mm. It is made of hardened steel.

TABLE 8.3 Physico-mechanical Properties (static)

ROCK	γ kN/m³	V_p km/s	V_s km/s	σ_{tBR} MPa	UCS MPa
Awaishi	22.4	4.43	2.49	3.4	28.1
Oya tuff	16.6	1.87	1.15	0.5–1.0	4.7–11.2

8.6.3 Experiments

8.6.3.1 Paraffin

Dynamic impact tests were carried out by varying drop height from 200 mm to 350 mm while keeping the weight of the drop cylinder by 5 kgf (49.1 N). Figure 8.30 shows views of the samples tested under static and dynamic conditions. The plastic zone in the vicinity of the impact object is whitish for the drop height of 200 mm, while radial tensile fractures (angle between fractures is 120 degrees) developed for the drop height of 350 mm in addition to whitish plastic zone.

Figure 8.31 shows the time-nominal tip pressure response for the drop heights of 200 and 350 mm of the impact object. Nominal tip pressure is defined as the ratio of dynamic force over the tip area of the cylindrical impact object.

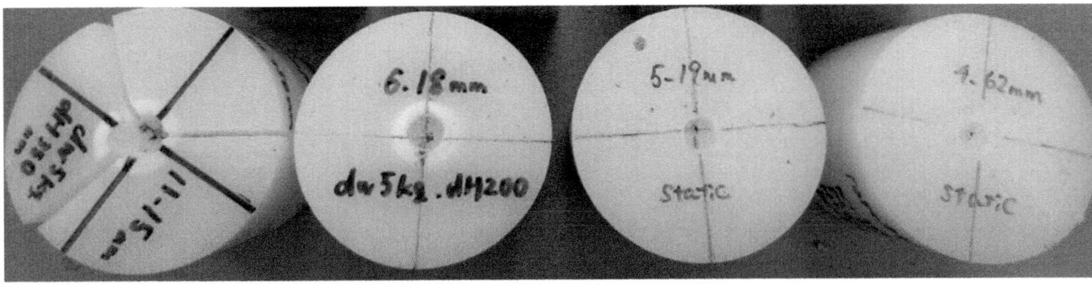

FIGURE 8.30 Views of samples tested under static and dynamic loading conditions.

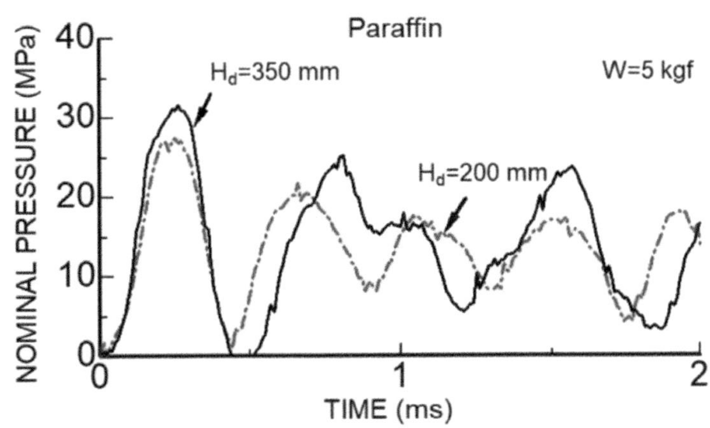

FIGURE 8.31 Time versus nominal tip pressure response for different drop height under a drop weight of 5 kgf.

8.6.3.2 Rocks

8.6.3.2.1 Experiments on Oya Tuff

Dynamic impact tests were carried out by varying drop height from 200 mm to 500 mm while keeping the weight of drop cylinder at 10 kgf (98.1 N). Figure 8.32 shows views of the samples tested under static and dynamic conditions. The plastic zones in the vicinity of the impact object appear as holes. In addition, radial fractures developed in the dynamic and static tests. Figure 8.33 shows the time-nominal tip pressure response for the drop heights of 200, 350 and 500 mm. Despite the variation of drop heights, the overall resistance remained almost constant.

FIGURE 8.32 Views of samples tested under static and dynamic loading conditions.

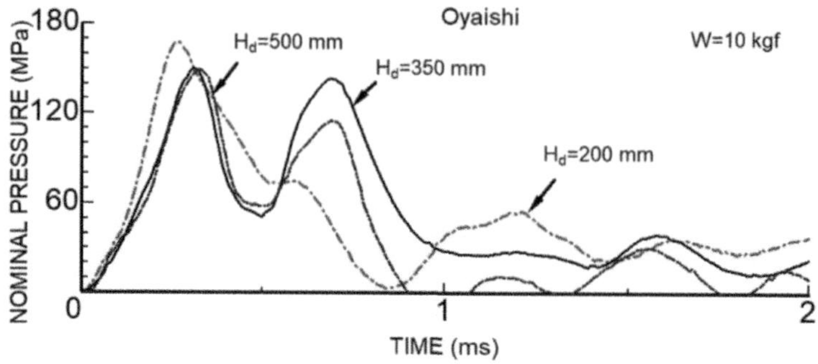

FIGURE 8.33 Time versus nominal tip pressure responses for different drop heights under a drop weight of 10 kgf.

8.6.3.2.2 Awaishi

Dynamic impact tests were carried out by varying drop height from 200 mm to 350 mm while keeping the weight of drop cylinder at 10 kgf (98.1 N). Figure 8.34 shows views of the samples tested under static and dynamic conditions. The plastic zones in the vicinity of the impact object appear as holes. In addition, radial fractures developed in dynamic and static tests. Figure 8.35 shows the time-nominal tip pressure response for the drop heights of 200 and 350 mm. Despite the variation of drop heights, the overall resistance remained almost constant.

FIGURE 8.34 Views of samples tested under dynamic loading conditions.

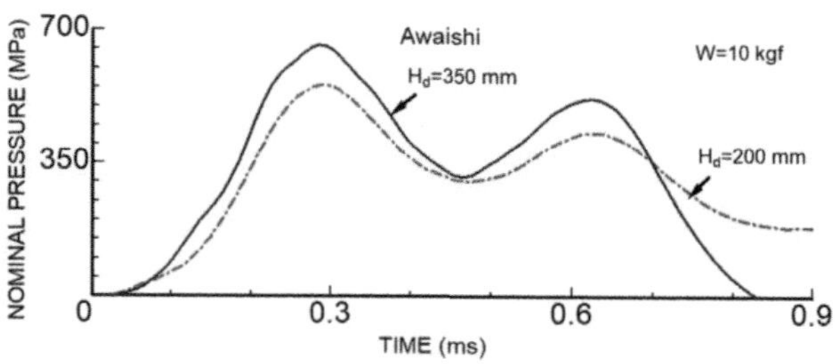

FIGURE 8.35 Time versus nominal tip pressure response for different drop heights under a drop weight of 10 kgf.

8.6.4 Evaluation of Experiments and Discussions

8.6.4.1 Effect of Momentum

As noted from the experimental setup, the height and weight of the drop object are the most important parameters on the amplitude of the imposed impact load. With this in mind, the ratio of measured dynamic static tip pressures is obtained from the experimental results plotted as in Figure 8.36, a function of the momentum theoretically defined by Eq. (8.21). The experimental results differ depending on the material type, and they are fitted to a linear function.

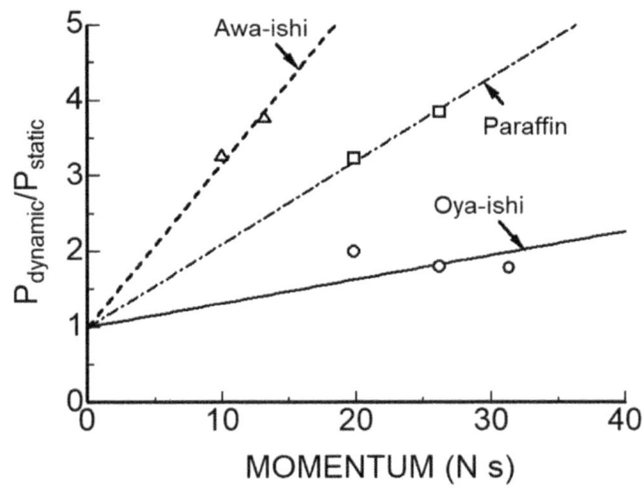

FIGURE 8.36 Momentum versus the ratio of dynamic tip pressure over the static tip pressure.

8.6.4.2 Penetration and Damage

The penetration depth and damage around the impact object differ depending upon the height and weight of the drop object as well as the characteristics of the materials. Figure 8.37 shows the penetration strain defined as the ratio of penetration (δ) over the diameter of the cylindrical impact object as a function of momentum given by Eq. (8.21). The experimental results are fitted to the following function:

$$\frac{\delta}{D} = aP^2 \tag{8.22}$$

where a is an empirical constant.

As noted from Figure 8.37, the empirical relation given by Eq. (8.4) presents a good fit to the experimental results. Figure 8.37 shows the plastic zone under dynamic conditions. The plastic zone development is co-centric and expands outward depending on the momentum imposed on the cylindrical impact object. Figure 8.38 shows the plastic zone development in the vertical cross section. As noted from the figure, a semispherical plastic zone occurs beneath the tip of the cylindrical impact object and propagates as the amount of the penetration increases. However, radial fractures develop as the penetration depth increases irrespective of material type.

8.6.4.3 Effect of Brittleness

As pointed out previously, paraffin is a perfectly plastic ductile material while Oyaishi and Awaishi are brittle materials. Figures 8.29, 8.31 and 8.33 show the post-test views of samples, while Figures 8.37 and 8.38 show close-up views of plastic zones. Fundamentally, plastic zone development is similar for both ductile and brittle materials. Nevertheless, radial fractures due to tensile stresses occur, and they become dominant after a certain amount of penetration.

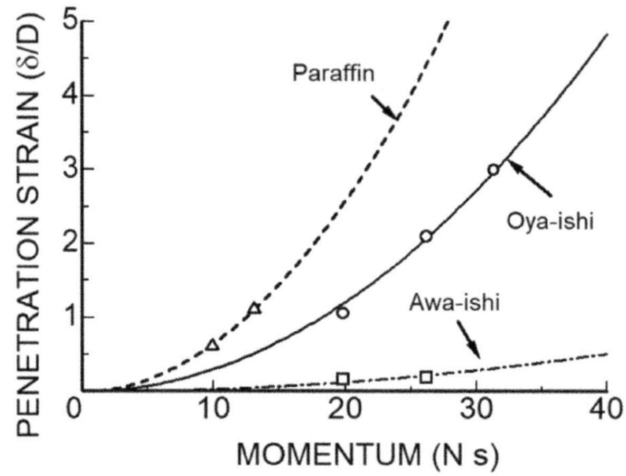

FIGURE 8.37 Experimental results for momentum versus penetrations strain.

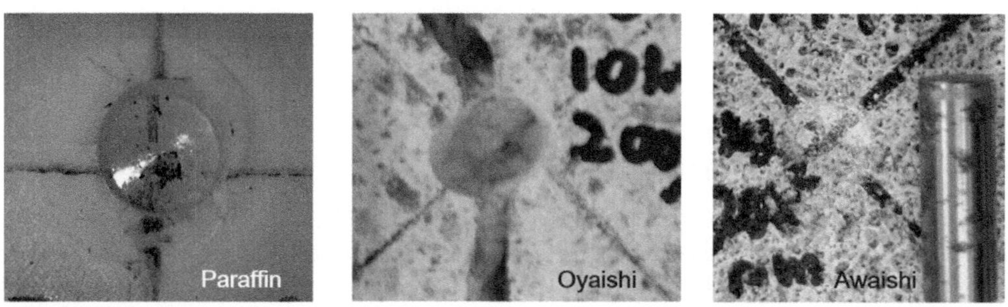

FIGURE 8.38 Plastic zones under dynamic conditions.

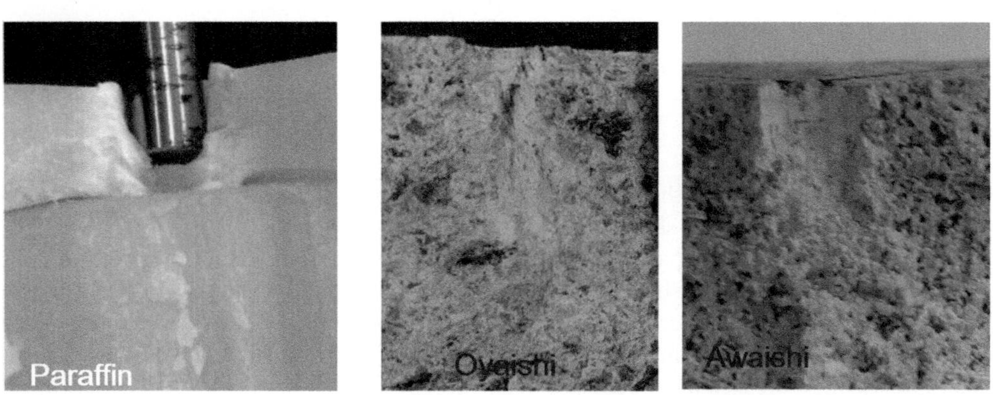

FIGURE 8.39 Plastic zones in vertical cross sections.

8.7 PENETRATION OF A PROJECTILE INTO ROCK BY BALLISTIC IMPACT

A penetration of a projectile (150 mm long and 25 mm diameter) with an initial velocity of 1650 mm/s into a sandstone block during a ballistic impact was investigated. The reaction side of the block was instrumented with two accelerometers capable of measuring three components simultaneously during impact and penetration of the projectile. Figure 8.40 shows the setup and instrumentation. Figure 8.41 shows the projectile penetration and reaction at several time steps. Figure 8.42 shows the cratering on the side where projectile hit the rock block and the close vicinity of the crater. Figure 8.43 shows the acceleration response and integrated displacement response of the sandstone block using the EPS method (Ohta and Aydan 2004, 2009).

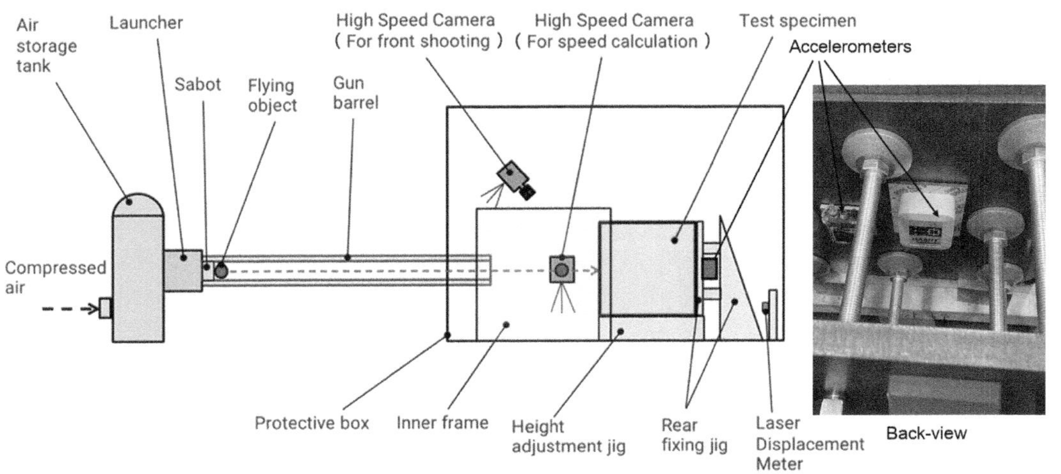

(a) Test chamber layout (b) Instrumentation on reaction side

FIGURE 8.40 Views of test setup and instrumentation (modified from Ohta et al. 2025).

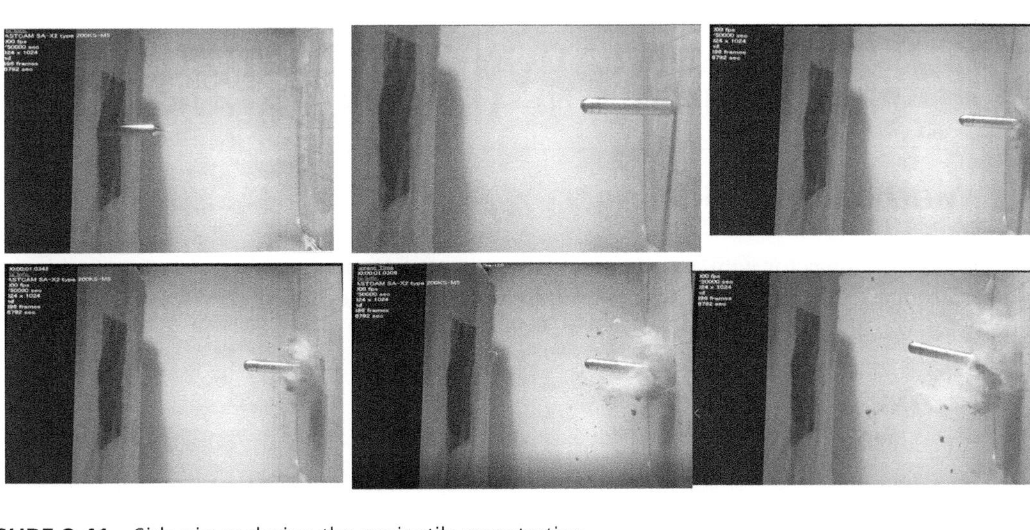

FIGURE 8.41 Side views during the projectile penetration.

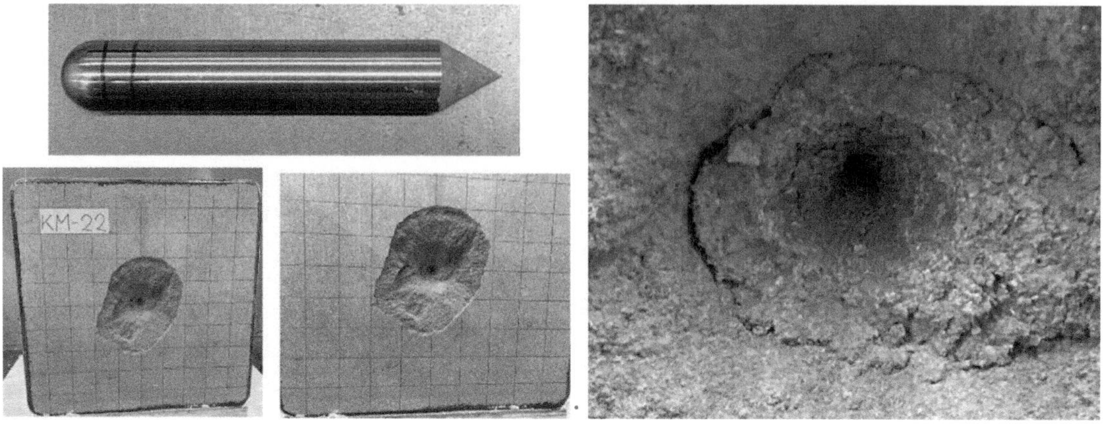

(a) Projectile and cratering (b) Close-up view of the crater

FIGURE 8.42 Views of the projectile and craters induced by the projectile.

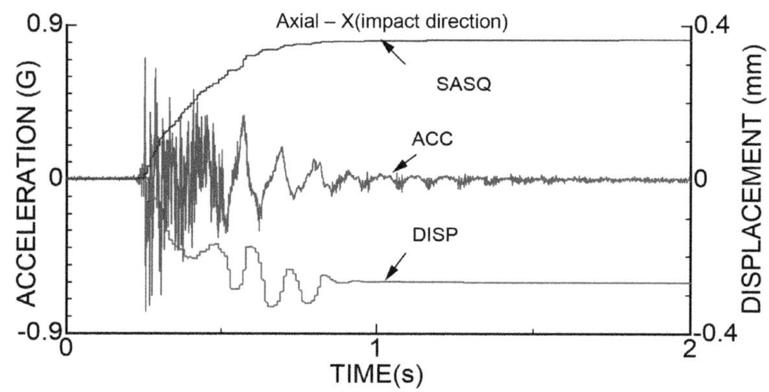

FIGURE 8.43 Acceleration response and integrated displacement response of the sandstone block using the EPS method (Ohta and Aydan 2004, 2009).

8.8 FORMULATION OF IMPACTOR PENETRATION AND ITS APPLICATIONS

8.8.1 Mechanical Modeling

Let us consider an impactor with a given mass (m) and geometry that is released at a given height H_f above the ground surface (Figure 8.44). If the air resistance is neglected, the dynamic equilibrium equation of the object during freefall may be written as follows:

$$\sum F_v = mg - m\frac{d^2 s}{dt^2} = 0 \tag{8.20}$$

where g is gravitational acceleration, and s is the distance from release point.

On the other hand, the resistance from ground against the motion of the object would be imposed following its impact and penetration into the ground, provided that the collision is inelastic (without rebounding). The resistance (F_r) of the ground may consist of three components, namely tip bearing resistance, friction between the object and ground and velocity dependent resistance. Therefore, one may write the following relation for the dynamic equilibrium equation of the object during impact and penetration:

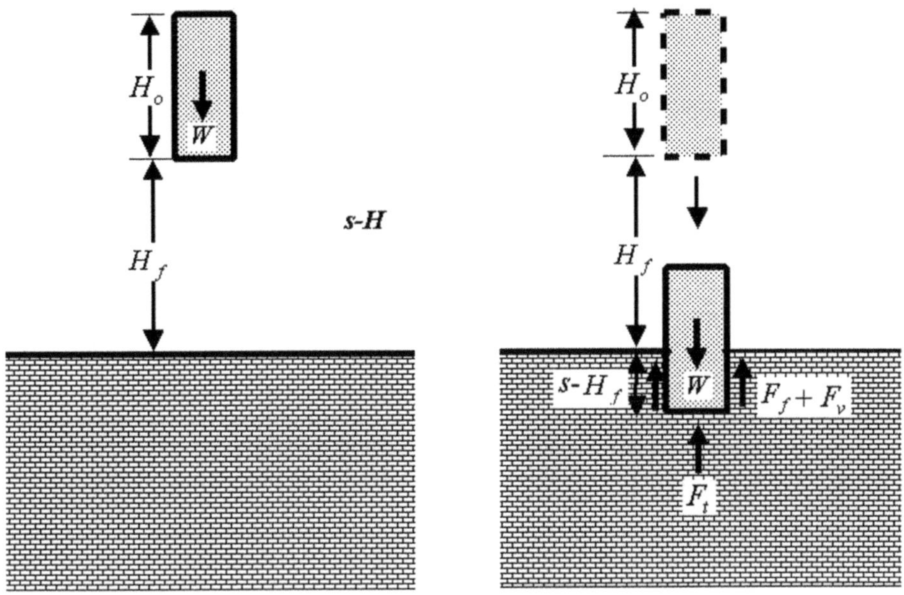

FIGURE 8.44 Mechanical modeling of penetration of impactor into ground.

$$\sum F_v = mg - F_r - m\frac{d^2 s}{dt^2} = 0 \tag{8.23}$$

The tip resistance of the ground may be evaluated using the concept of the bearing capacity of foundations and proportional to the mass of the falling object as follows ($s>H_f$):

$$F_t = mg \cdot q(1 + \alpha \frac{s - H_f}{D}) \tag{8.24}$$

where

$$q = \frac{1 + \sin\phi}{1 - \sin\phi} \cdot \frac{\rho_s}{\rho_o}; \ \alpha = \tan(45 - \frac{\phi}{2}) \tag{8.25}$$

The tip resistance may be neglected for sufficiently slender objects. The second term on the right-hand side of Eq. (8.24) takes into account the effect overburden following the penetration of the object.

The frictional resistance between the object and surrounding ground may be given as follows:

$$F_f = \mu \frac{s - H_f}{H_o} \text{mg for } (s - H_f) < H_o \tag{8.26}$$

and

$$F_f = \mu mg \text{ for } (s - H_f) > H_o \tag{8.27}$$

The viscous component of the resistance of ground against penetration may also be given in the following form:

$$F_V = mg \cdot \beta v^n \tag{8.28}$$

where β is viscous coefficient, v is velocity and n is a power coefficient. Thus the total resistance from the ground against impact and penetration would be a sum of these components as follows:

$$F_r = F_t + F_f + F_v \tag{8.29}$$

8.8.2 Solution Procedure

While Eq. (8.23) is quite easy to solve, Eq. (8.29) would be difficult to solve due to its nonlinear characteristics of viscous resistance and cumbersome conditions resulting from the tip bearing resistance and frictional resistance. Therefore, a numerical integration technique would be employed in this study. For a given time, step $n + 1$, Eqs. (8.23) and (8.29) may be represented in a unified style:

$$\left(\frac{d^2 s}{dt^2}\right)_{n+1} = A_r^{n+1} \tag{8.30}$$

with initial conditions:

$$s = 0 \text{ and } v = 0 \text{ at } t = 0 \tag{8.31}$$

The velocity and distance from the release point can be obtained from the following discretized equation as follows, using the Feynman–Newton method:

$$v_{n+1} = v_n + A_r^{n+1} \Delta t \tag{8.32a}$$

$$s_{n+1} = s_n + v_{n+1} \Delta t \tag{8.32b}$$

with

$$v_0 = \frac{1}{2} A_r^1 \Delta t \tag{8.32c}$$

8.8.3 Examples

Some computational examples are given herein, and their implications are discussed. The physical and geometrical properties of ground, object and freefall height are given in each respective figure. In computations, three different cases are analyzed:

CASE 1: Tip resistance only
CASE 2: Tip and side frictional resistance
CASE 3: Tip, side friction and viscous resistance

Figure 8.45 shows the computed acceleration, velocity and displacement responses for the properties shown in the same figure. In the freefall stage, the acceleration is constant, velocity increases linearly as expected from the integration of Eq. (8.23). After impact, the object starts to penetrate into the ground. As a result, the velocity starts to decrease, and the motion of the impactor tends to decelerate. When the velocity of the impactor becomes nil, the motion of the impactor terminates.

The computational responses for CASE 2 and CASE 3 are fundamentally similar to those for CASE 1 as shown in Figures 8.46 and 8.47. The biggest difference is that the penetration depth decreases due to the increase of the resistance from the ground.

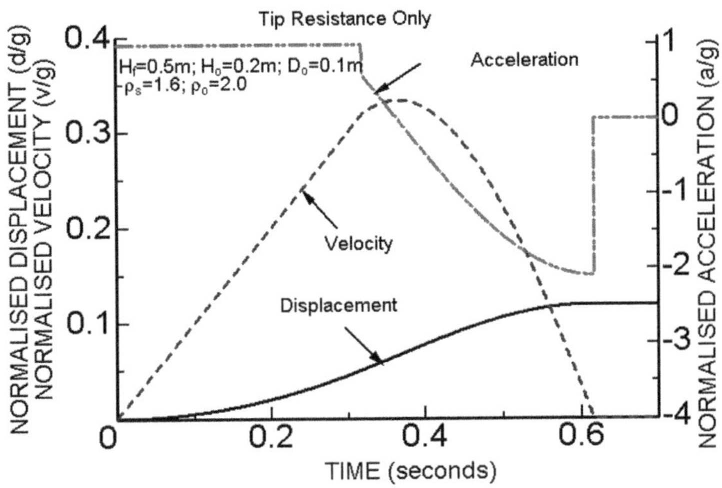

FIGURE 8.45 Computed responses of the acceleration, velocity and penetration depth of an impactor (CASE 1: Tip resistance only).

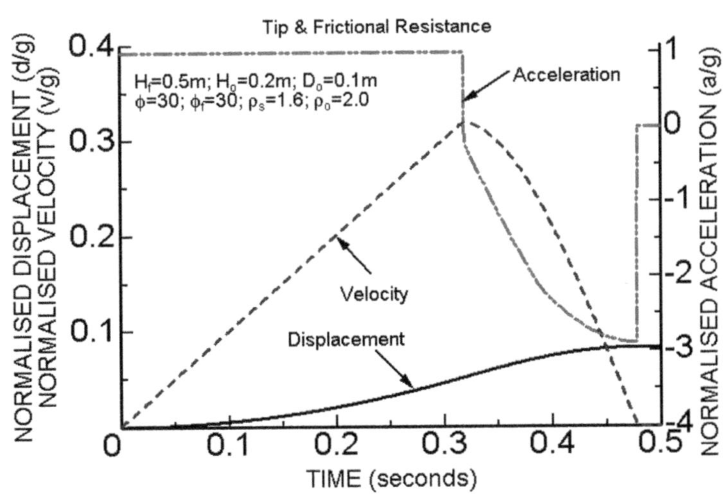

FIGURE 8.46 Computed responses of the acceleration, velocity and penetration depth of an impactor (CASE 2: Tip + frictional resistance).

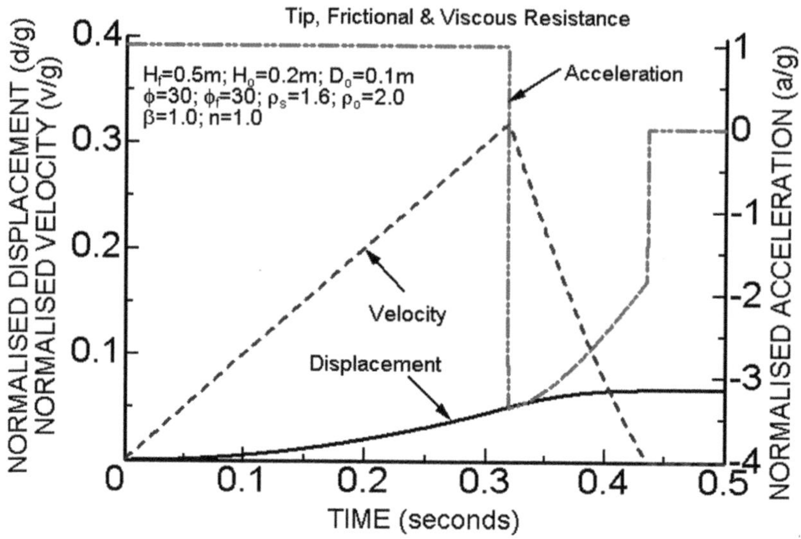

FIGURE 8.47 Computed responses of the acceleration, velocity and penetration depth of an impactor (CASE 3: Tip + frictional and viscous resistance).

8.9 PAST AND RECENT METEORITE IMPACTS ON MARS

Some 385000 impact craters having a diameter greater than 1 km exist on Mars. The largest one, named Huygens, has a diameter of 467.25 km, and 20 craters have diameters greater than 289 km. Curiosity MER has been exploring Gale Crater since August 6, 2012, while Perseverance MER has been exploring Jezero Crater since July 30, 2020. Besides MERs, Mars Reconnaissance Orbiter (MRO) provides many images of meteorite impacts. The InSight equipped with seismometers also recorded vibrations caused by the meteorite impacts. Figures 8.3 and 8.48 show several examples of recent impacts on Mars. Following the impacts, the ground is partially ejected, and it looks like dark material displaced. Although it may be contemplated that some fluid-like material is ejected, it may be such that the ejected material may be non-weathered material beneath the ground surface.

8.10 METEORITE IMPACTS AND CRATERS ON THE MOON

On the Moon. the total number of craters having a diameter of 1 km is 1.3 million. The largest crater has a diameter of 290 km and is located near the south pole. Craters constitute 95% of lunar features. The walls of some of these craters are unstable, and the crater walls are collapsing, as seen in Figure 8.49. Figure 8.50 shows the impact effects of used spacecrafts of NASA on the Moon.

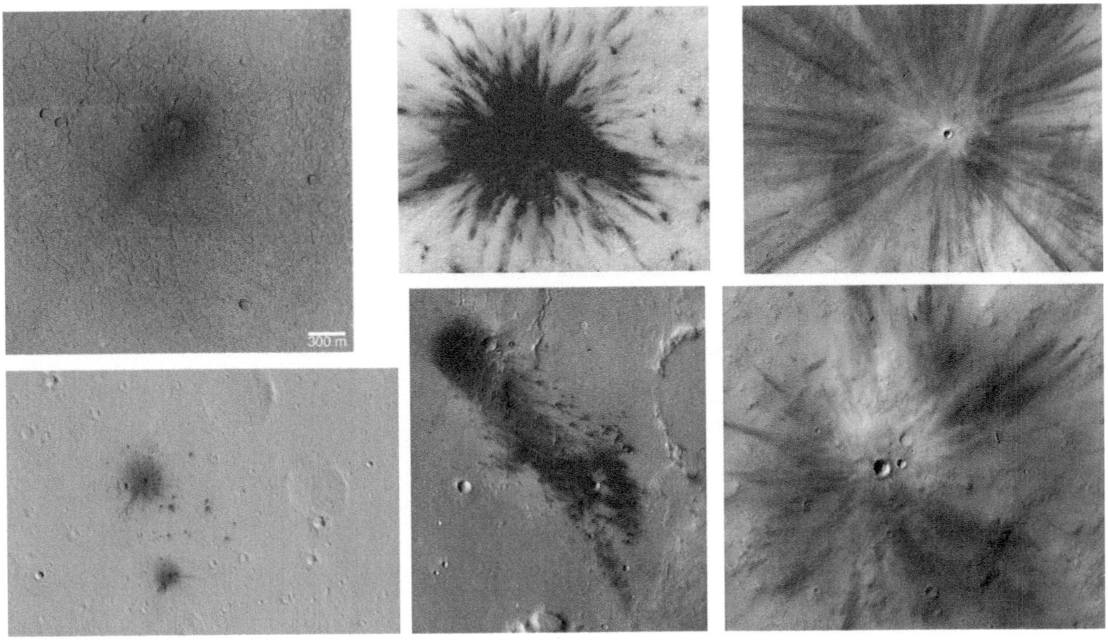

FIGURE 8.48 Some recent meteorite impacts in Mars (from NASA).

FIGURE 8.49 Ohm Crater with collapsed crater rim.

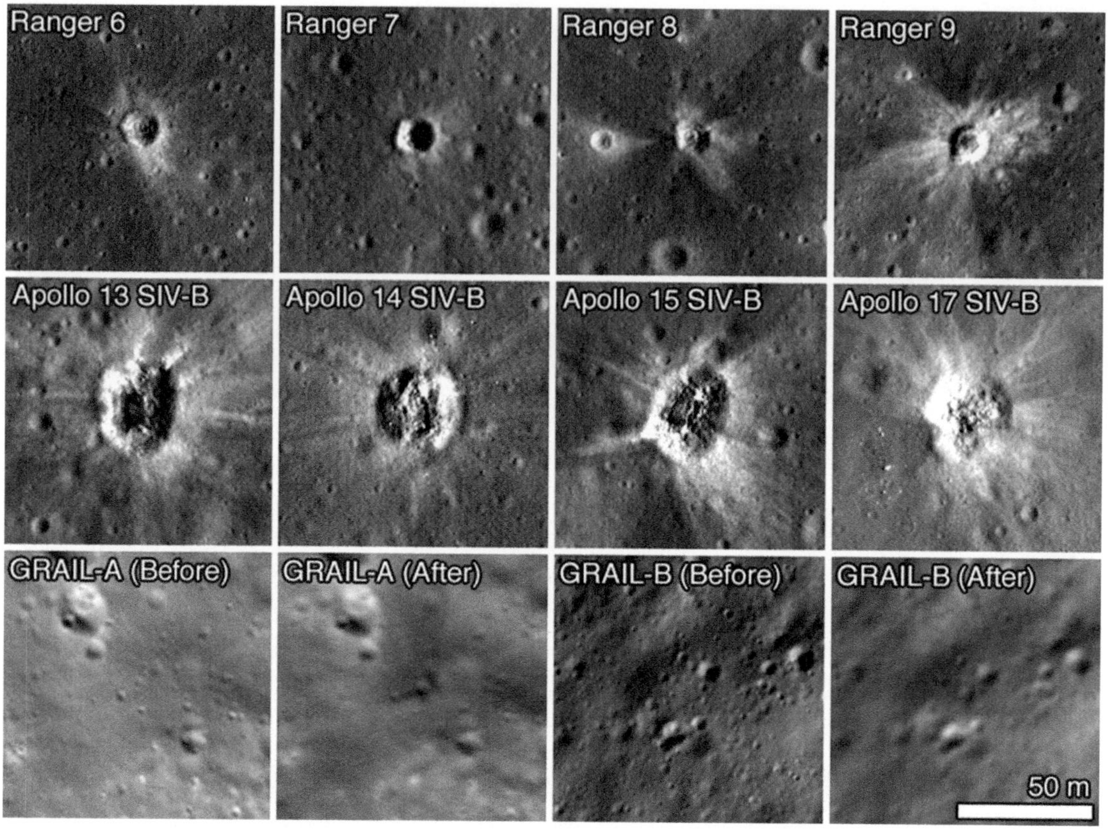

FIGURE 8.50 Impact effects of used spacecraft on the Moon (from NASA).

8.11 METEORITE IMPACTS AND CRATERS ON VENUS

As stated previously, direct observation on Venus is impossible due to thick clouds. However, the Magellan spacecraft explored the surface of Venus. and about 900 craters have been identified so far. The largest crater is named Mead, having a diameter of 270 km. Figure 8.51 shows a Magellan radar image of Mona Lisa crater with a diameter of 79.4 km. Although smaller size craters less than 300 m in diameter might exist on Venus, it is contemplated that the craters having a diameter less than 1.5–2 km may not exist due to the resistance of the thick atmosphere of Venus and the disintegration of meteoroids before impact.

8.12 METEORITE IMPACTS AND CRATERS ON MERCURY

As of 2021, the number of craters on Mercury is 414. The largest crater is named Caloris with a diameter of 1525 km. Figure 8.52 shows a mosaic image obtained when both the spacecraft and the Sun were overhead. Caloris basin has been flooded by lava. Craters cover much of the surface of Mercury, which are its common topographic features. Craters at the poles are presumed to be filled with ice due to permanent shadowing.

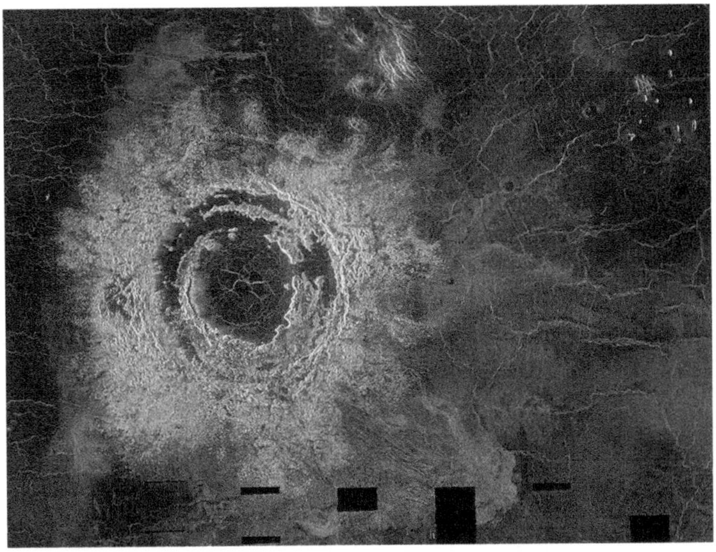

FIGURE 8.51 Magellan radar image of Mona Lisa crater on Venus (from NASA).

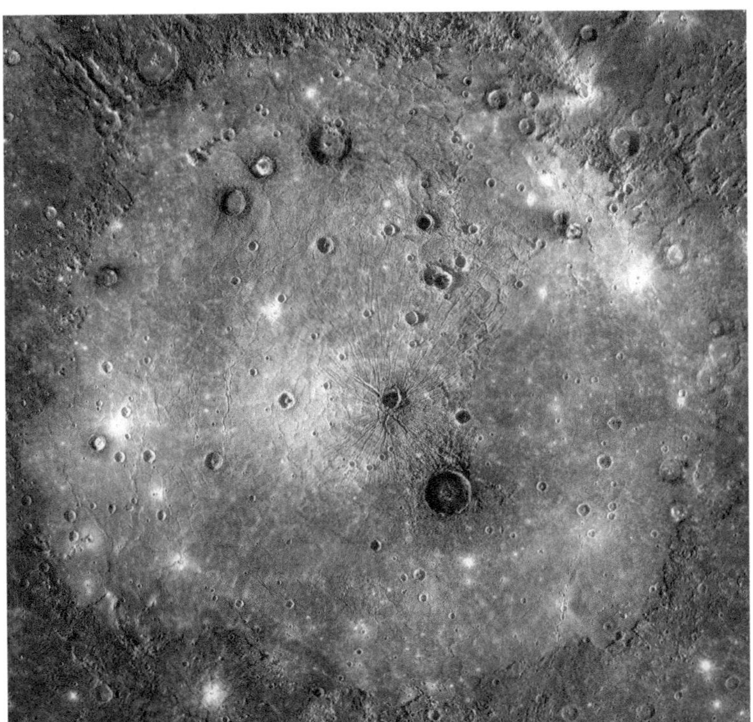

FIGURE 8.52 Image of the largest crater on Mercury, Caloris (from NASA).

Slope Stability

9

In this chapter, rock slope stability issues on some terrestrial planets and the Moon are outlined through the images provided by NASA. Rock slope stability issues are investigated in the context of the discontinuous nature of rock mass, the geometry of slopes, differential weathering, the effects of quakes, impacts by meteorites, and knowledge gained on rock slope stability issues on Earth. Some techniques developed for slopes on Earth were applied to slope failures observed in other terrestrial planes, and the outcomes are presented and discussed (Aydan 1989, 2007, 2009, 2015, 2020b, 2023a; Aydan and Kawamoto 1992, Aydan and Kumsar 2010; Aydan and Ulusay 2002; Aydan et al. 1989, 1991, 1992a, 1992b, 1996c, 1996d, 2005a, 2005b, 2007a, 2007b; Kumsar et al. 2000; Tokashiki and Aydan 2010).

9.1 MARS

The images of capturing and drilling operations performed by Mars Exploration Rovers (Opportunity, Spirit, Curiosity, Perseverance) on some selected rocks and rock blocks have indicated that the rocks are quite similar to those of Earth. Whether the rocks seen on Earth fully exist on Mars still needs some clarification. However, the sedimentary rocks (such as conglomerate, sandstone, siltstone, mudstone, sulfate deposits), metamorphic rocks such as phyllite, shale and serpentine and extrusive rocks (basalt, andesite-like (silicate-bearing basalt), breccia) of Mars are clearly similar to those of Earth. Although the present climatic conditions without rainfalls and the nonexistence of groundwater and large difference in daily temperature, the failures due to gravitation and seismic forces, impacts, differential weathering are likely to occur. Furthermore, it is well understood that the circumstances in the past were quite similar to those of Earth and many paleo rock slope failures should be very similar to those observed on Earth. In this section, rock slope failures in Mars, on the bases of the images provided by Mars Rovers of NASA, are summarized with consideration of the following conditions:

1. Slope failures in layered rock mass
2. Slope failures in jointed rock mass
3. Actual or potential slope failures in discontinuous rock mass
4. Steep or overhanging cliffs in layered rock mass
5. Failures of overhanging cliffs in discontinuous rock mass
6. Stability of precarious rock blocks
7. Impact and vibration-induced mass movements

Aydan (1989) categorized and classified rock slope failures on Earth as shown in Figure 9.1. Fundamentally, rock slope failures involve three categories, specifically, intact rock or rock mass, discontinuities, and intact rock and discontinuities. They are called shearing or bending failure, combined shearing and sliding, buckling, flexural toppling, planar and/or wedge sliding and columnar toppling.

DOI: 10.1201/9781003643852-9

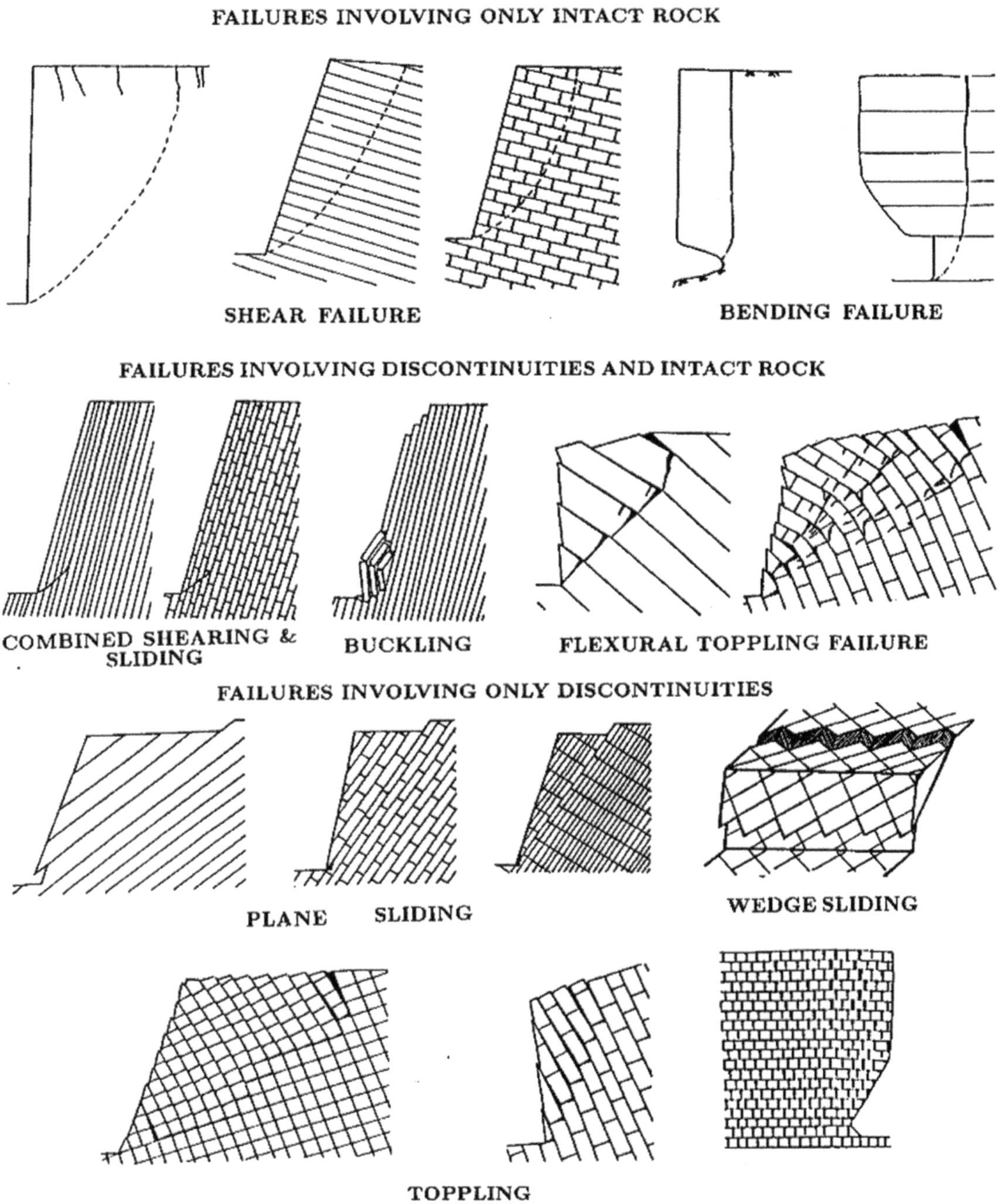

FIGURE 9.1 Classification of rock slope failures (from Aydan 1989).

9.1.1 Curved Shear Slope Failures

It is reported that the large tectonic structure known as Valles Marineris, which is 4000 km long, 200 km wide and 7000 m deep, involves many curved shear failures of very large scale. The stability of slopes in Valles Marineris in Mars was initially investigated by several pioneering researchers (e.g., Brunetti 2014; Conway et al. 2011; Crosta et al. 2014; Lucchitta 1979; Neuffer and Schultz 2006; Schultz 2002). Figure 9.2 illustrates some curved shear slope failures of large scale; the height of the failed slopes is several thousand meters.

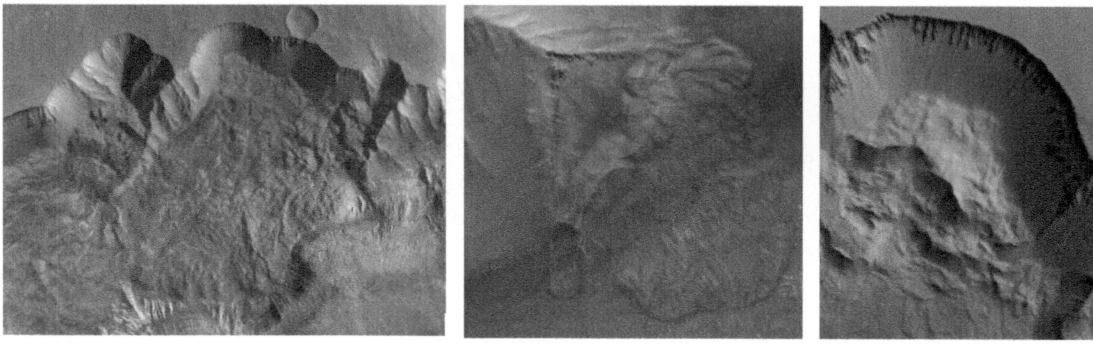

FIGURE 9.2 Large-scale curved shear failures.

9.1.2 Bending Failures

Bending failure is generally observed when the toe of the slope is eroded due to differential weathering resulting from environmental conditions and fluid flow. Figure 9.3 shows several examples from Mars. As noted from the images, some layers failed due to bending failure, and the failure was restrained by some layers with higher resistance.

FIGURE 9.3 Examples of bending failure of layered rock slopes.

9.1.3 Flexural Toppling Failures

Flexural toppling failure occurs when the rock mass is layered (Aydan and Kawamoto 1992). The images from Mars rovers revealed very thinly layered rock masses that are prone to flexural toppling, and such failures were observed as seen in Figure 9.4; rock masses were mainly mudstone, shale and sandstone. In some cases, conglomeratic layers failed in flexural toppling mode.

FIGURE 9.4 Examples of flexural toppling failures.

9.1.4 Planar and Wedge Failures

While planar failures involve one planar surface, wedge failures involve two crossing planes, whose lines of intersection are inclined out of slopes. Figure 9.5 shows some typical examples of planar and wedge failures observed along the route of Curiosity Mars Rover in Gale Crater. In addition, some planar sliding failures involving two planes inclined out of slopes are noted. In such cases, the major failure plane is the bedding failure while the other plane are cross-joints. Besides the gravitational forces, dynamic loads resulting from impacts of meteorites or mars quakes could have been triggering factors.

FIGURE 9.5 Examples of planar and wedge failures.

9.1.5 Block/Columnar Toppling Failures

Block/columnar toppling failure may occur due to rotational motion under certain force conditions. Toppling could be due to the rotation of a column or columns consisting of a single block or multiple blocks (Aydan et al. 1989). Figure 9.6 shows several examples of such failures.

FIGURE 9.6 Examples of block/columnar toppling failures.

9.1.6 Rockfalls

Some very recent large scale rockfalls were reported by NASA through high-resolution images taken by the Mars Reconnaissance Orbiter (MRO). These images clearly showed even the paths of rockfalls, which generally occurred along rift zones and steep cliffs. The MRO also recorded a rockfall event, as seen in Figure 9.7.

FIGURE 9.7 Rockfalls (rock avalanche) events in Mars.

9.1.7 Precarious (Balancing) Rocks

On Earth, many precarious rock blocks are shaped due to differential weathering, winds and fluid flow in addition to natural discontinuities. As seen in Figure 9.8, the images from Mars Exploration Rovers (MERs) showed very interesting precarious (balancing) rocks. Such rocks receive great interest among scientists and engineers involved with the inference of the seismic motions of the ground and structures in non-instrumented areas particularly. Undoubtedly, the evaluation of precarious rocks observed by the MERs may be of great value for investigating vibrations caused by meteorite impact, marsquakes and even the MERs themselves.

FIGURE 9.8 Views of precarious rocks.

9.1.8 Evaluation of Rock Slope Failures in Mars, the Moon and Venus

Aydan et al. (1991) proposed an integrated rock slope stability assessment system, which consider different failure modes of rock slopes and provide their most likely stability limits. This system considers seismic forces in addition to gravitational and groundwater forces. Figure 9.9 shows an application of the integrated slope stability assessment to rock slopes of Mars, the Moon and Venus, together with observations denoted by different symbols. Lines in the figure represent the relations between the bedding plane and lower slope angle for the stability of rock slopes for jointed rock mass. The friction angles used in this graph are those of basaltic rock joints on Earth under dry conditions.

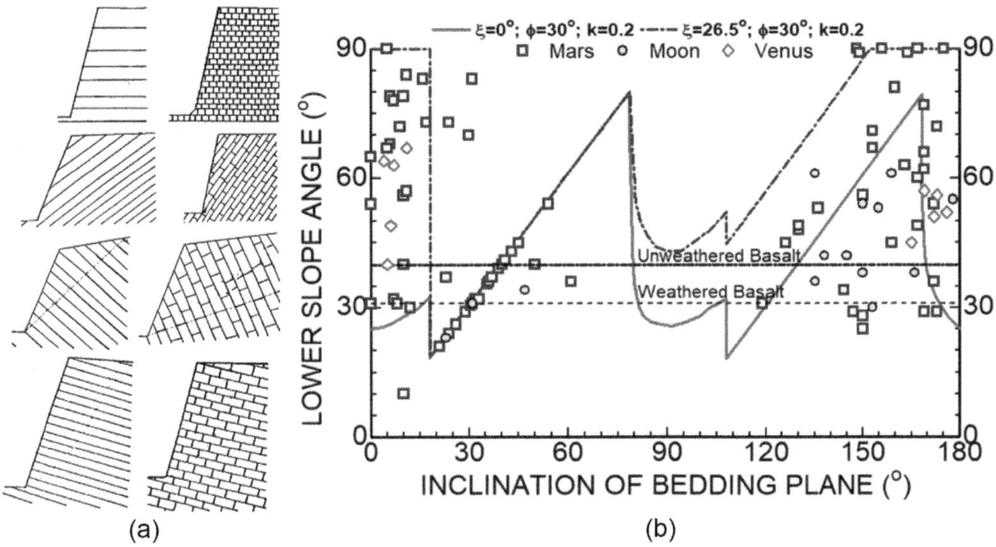

FIGURE 9.9 (a) Illustration rock mass conditions and (b) the relation between the inclination of the bedding plane and lower slope angle for rock slopes on Mars, the Moon and Venus.

9.1.9 Vibration-induced Mass Movements

Meteorites of various sizes bombard planets, including the Earth and Moon, and quakes (earthquake, moonquake, marsquake) occur from time to time. A seismic record by the seismometer of the InSight Mars module confirmed a marsquake on April 24, 2019, for the first time as presumed by Anderson et al. (1977). These events result in shock waves and vibrations. Figure 9.10 shows several examples of recent impacts and the displaced rock blocks in Mars, and Figure 9.11 shows the slope failures in the vicinity of a small impact crater.

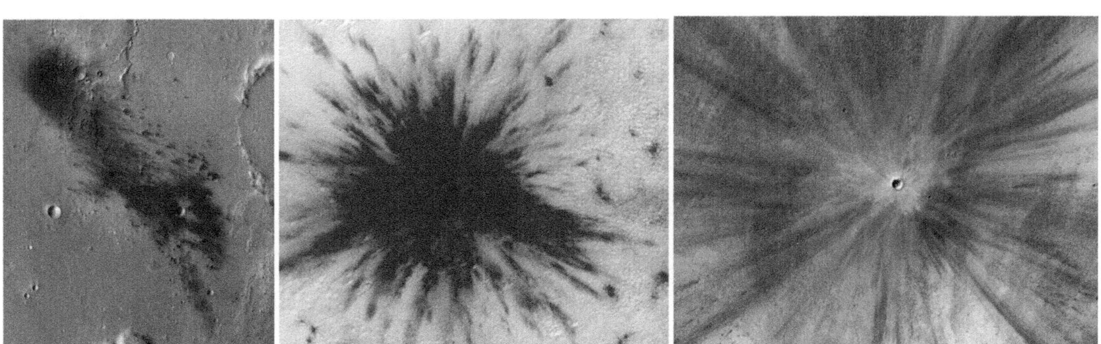

FIGURE 9.10 Views of recent meteorite impacts in Mars.

FIGURE 9.11 Meteorite impact–induced slope movements in the vicinity of a small-scale impact crater.

A rock block displaced on Mars, as shown in Figure 9.12, is considered and the conditions for its motion are analyzed next. The travel path length was 675 cm, and the inclination of the path was almost 23.5 degrees. As rocks at the site were inferred to be basalt, the result of a dynamic friction experiment on a saw-cut discontinuity of basalt from Mt. Fuji (Aydan 2019, 2020a) was utilized (Figure 9.13). The simple mechanical considerations shown in Figure 9.14 yielded the maximum acceleration and velocity to the displacement of the rock block (Aydan 2020c, 2021a, 2022a, 2022b):

$$\text{Maximum Acceleration: } a > g_m \tan(\phi - \alpha) \tag{9.1}$$

$$\text{Maximum Velocity: } v_{\max} > \sqrt{2g_m \tan(\phi - \alpha)\delta} \tag{9.2}$$

FIGURE 9.12 Displaced rock block.

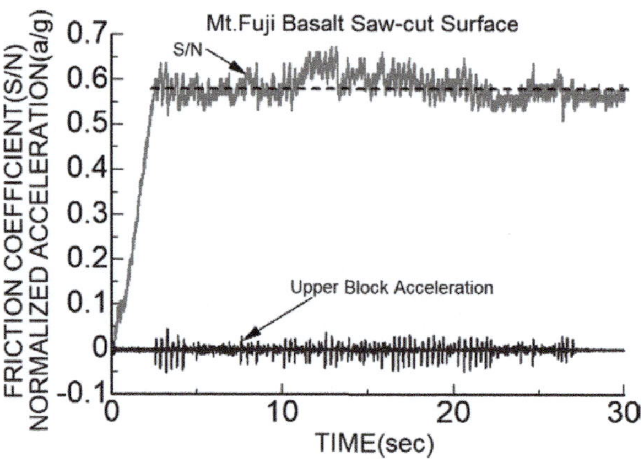

FIGURE 9.13 Dynamic friction test on a basaltic saw-cut discontinuity.

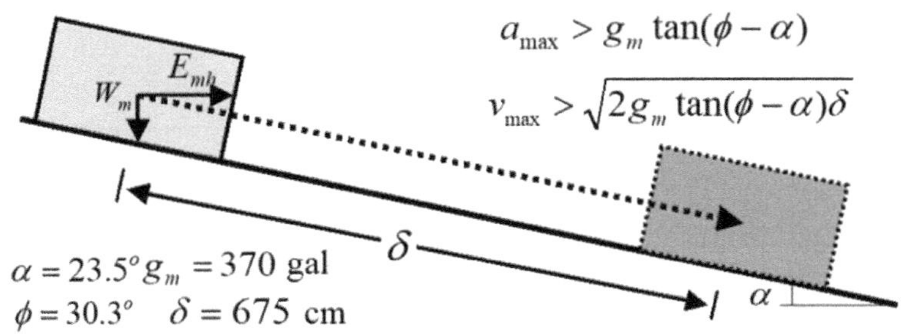

FIGURE 9.14 Mechanical model of rock block subjected to gravity and impact forces.

As seen from Figure 9.13, the friction angle of the basaltic saw-cut surface from Mt. Fuji is greater than 30 degrees. If these values are used, the conditions to displace the rock block shown in Figure 9.12 are estimated to be:

Maximum Acceleration: 42.13 cm/s^2
Maximum Velocity: 238.5 cm/s

The magnitude of the earthquake for this maximum ground acceleration value can be estimated to be about 3.8–3.9 using some empirical relations developed for earthquakes on Earth (Aydan 2012a). As understood from this simple example, rock dynamics would be necessary for the assessment of various rock mechanics and rock engineering issues on Mars and other planets.

9.2 MOON

The slope stability issues exist, and rockfalls still continue since the formation of the Moon (Figures 9.15 and 9.16). Bickel et al. (2021) studied a huge number of high-resolution images of the Moon and concluded that the main driving causes of mass movements, rockfalls and erosion on the Moon in its airless environment are the vibrations resulting from moonquakes, impacts by meteorites and extreme temperature variations. Furthermore, the sampling at Taurus-Littrow Valley near the Apollo 17 landing site provided geological conditions at the mass-wasting and rockfall site. Magnarini et al. (2023) performed direct shear tests on granulated anorthosite samples under different conditions, such as dry or humid, which were subjected to normal stress levels ranging between 2 and 10 MPa under atmospheric pressure and vacuum condition, which showed that peak friction angle ranges between 30 and 33 degrees and the residual friction angle between 26 and 30 degrees. Furthermore, the stick-slip behavior becomes quite dominant as the normal stress level increases. The maximum normal stress level of the tests would correspond to a depth of 2300 m from the ground surface under lunar conditions.

9.3 VENUS

Direct optic detection of mass movements, slope failures and rockfalls is not possible due to the thick cloud layer enveloping Venus. Therefore, any detection is possible only through radar imaging. Malin

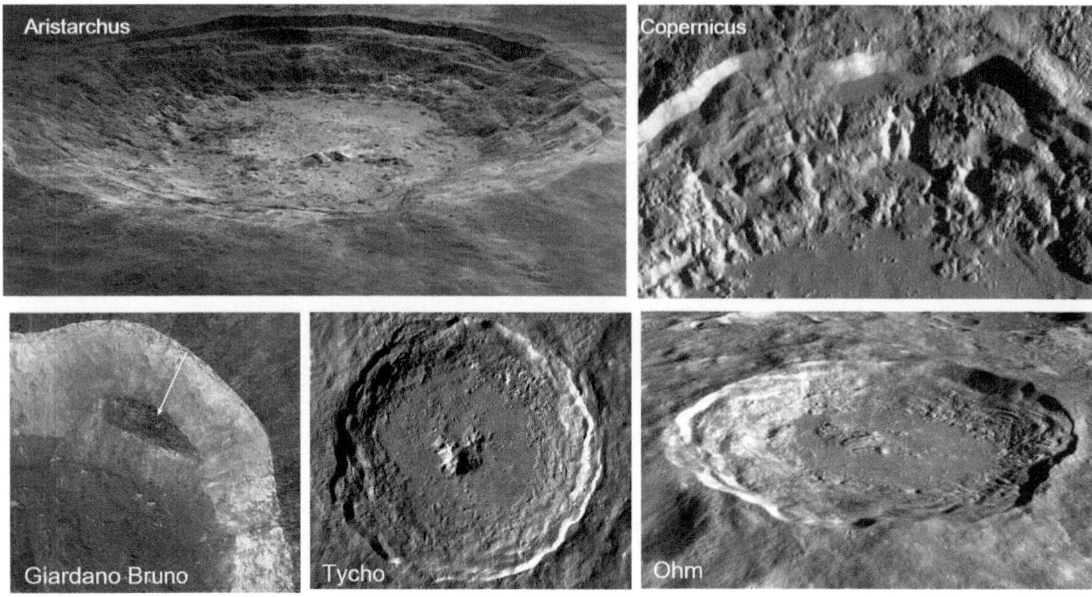

FIGURE 9.15 Large-scale circular-sliding failures in lunar craters.

FIGURE 9.16 Precarious rock blocks and rockfalls on the Moon.

(1992) reported mass movements on Venus seen in radar images acquired by the Magellan spacecraft during its first mapping cycle. Malin (1992) could interpret rock slumps, rock and/or block slides, rock avalanches, debris avalanches, and possibly debris flows. They are seen in areas of high relief and steep slope gradients, and they are most abundant in the tectonic troughs that crisscross much of the equatorial region of Venus. Events smaller than the resolution of 75 m could not be identified. Malin (1992) could also identify mass movements in volcanic areas. However, further radar-based imaging of Venus is necessary for more detailed interpretations of mass movements in Venus.

9.4 MERCURY

As Mercury is very close to the Sun, topographic mapping was not possible until 1990. The mapping was achieved by the Mercury Laser Altimeter and Mercury Dual Imaging System on the Messenger spacecraft. Anything smaller than a resolution of 17–36 m could not be detected. Nevertheless, several researchers attempted to interpret these images to detect mass movements such as rock slumps, rock and/or block slides, rock avalanches, debris avalanches, debris flows and impact ejecta. It was also possible to identify mass movements in volcanic areas. There are also some attempts to evaluate rockfall events, which require much higher resolutions (Kreslavsky et al. 2021). Figures 9.17 and 9.18 show several examples of evaluations. No doubt mass movements in Mercury should continue currently in view of what happens on other terrestrial planets if continuous-type reconnaissance orbiters can be utilized for Mercury as they have been for Earth, Mars and the Moon.

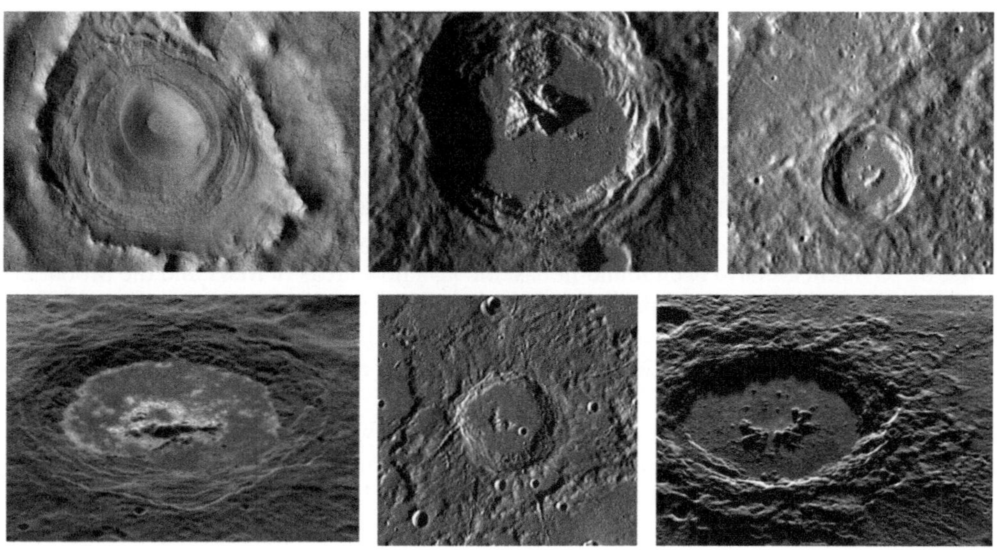

FIGURE 9.17 Mass movements in some of Mercury craters.

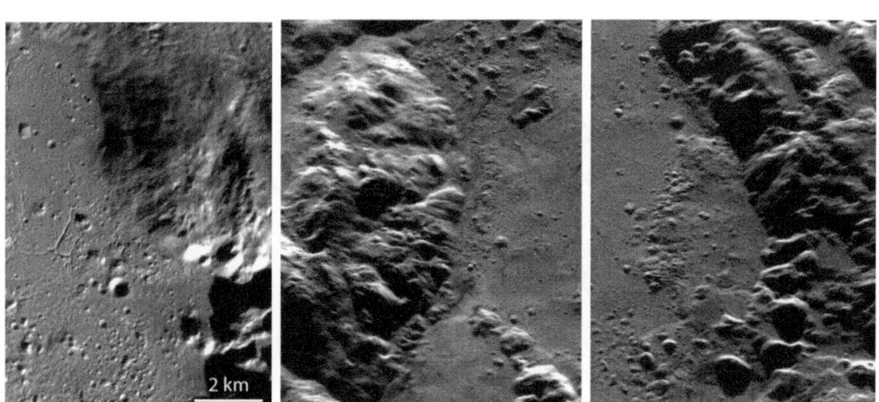

FIGURE 9.18 Views of rockfalls in Mercury craters.

Stability of Underground Cavities and Sinkholes (Skylights) 10

Sinkholes named in Geoengineering and Skylights or pits in Planetary Astronomy cause severe problems on the Earth (e.g., Cruikshank and Wood 1971; Cushing 2012; Cushing et al. 2007, 2015; Haruyama et al. 2012; Wagner and Robinson 2014). This problem has been often observed in karstic terrains, evaporitic formations and volcanic regions on Earth (e.g., Aydan 2008, 2018b, 2021a; Aydan et al. 2017; Aydan and Kumsar 2005; Aydan and Tokashiki 2007, 2011; Aydan and Kawamoto 1992; Aydan and Ulusay 2003, 2013; Barton 1976; Barton et al. 1980; Kawamoto et al. 1991; Kumsar and Aydan 2022; Tokashiki 2011). The preliminary observations by the Mars and lunar rovers have indicated that these sinkholes have been found in basaltic lava flow areas in Mars and the Moon. Some of sinkholes may also be formed by impacts of meteorites and/or rifting induced by tectonic movements. Radiation is a major issue for manned-space explorations, and the sinkholes of Mars and the Moon may provide a shelter against radiation. Therefore, their stability is of great importance for their habitation and utilization.

In this chapter, the underground cavities and sinkhole (skylight) issue on Mars and the Moon are considered with some references to some of the sinkholes of Earth, such as the volcanic sinkholes, and caves in Japan, Türkiye, Hawaiian Islands and mainland of the US. The geometry and size of sinkholes of Mars and the Moon are categorized, and their stability is evaluated on the basis of Rock Mechanics and Rock Engineering and then compared with those of Earth. The stability evaluations involve empirical rock classifications, analytical methods and numerical methods. Furthermore, the possibility of similar underground structures on Mercury and Venus are discussed.

10.1 SINKHOLES (SKYLIGHTS) AND CHARACTERISTICS OF VOLCANIC ROCKS ON MARS AND THE MOON

10.1.1 Sinkholes (Skylights)

Sinkholes may form due to the collapse of natural underground openings under gravity, seismic loading, groundwater seepage and degradation of rocks. This problem has been often observed in karstic terrains, evaporitic formations and volcanic regions on Earth. The flow of lava in volcanic terrain under gravity may form underground structures named lava tubes due to rapid cooling and hardening of lavas

DOI: 10.1201/9781003643852-10

at the ground surface. As reported by Cushing (2012) and Cushing et al. (2007, 2015), a number of sinkholes, mainly in the basaltic lava flow plains of Mars, can be seen in Figure 10.1. In these images, one can also notice sinkholes (skylights), which are used to infer the geometry of sinkholes such as depth, size and cover rock thickness. In the same figure, one can also notice the natural rock bridges, which could be interpreted as further evidence for the cause of the sinkhole formation due to the collapse of the rock cover. Undoubtedly, sinkholes may also be formed due to the impacts of meteorites and/or rifting induced by tectonic movements. Although there are presently no water-like fluid bodies in sinkholes, they may have been associated with their formation in Mars in the past.

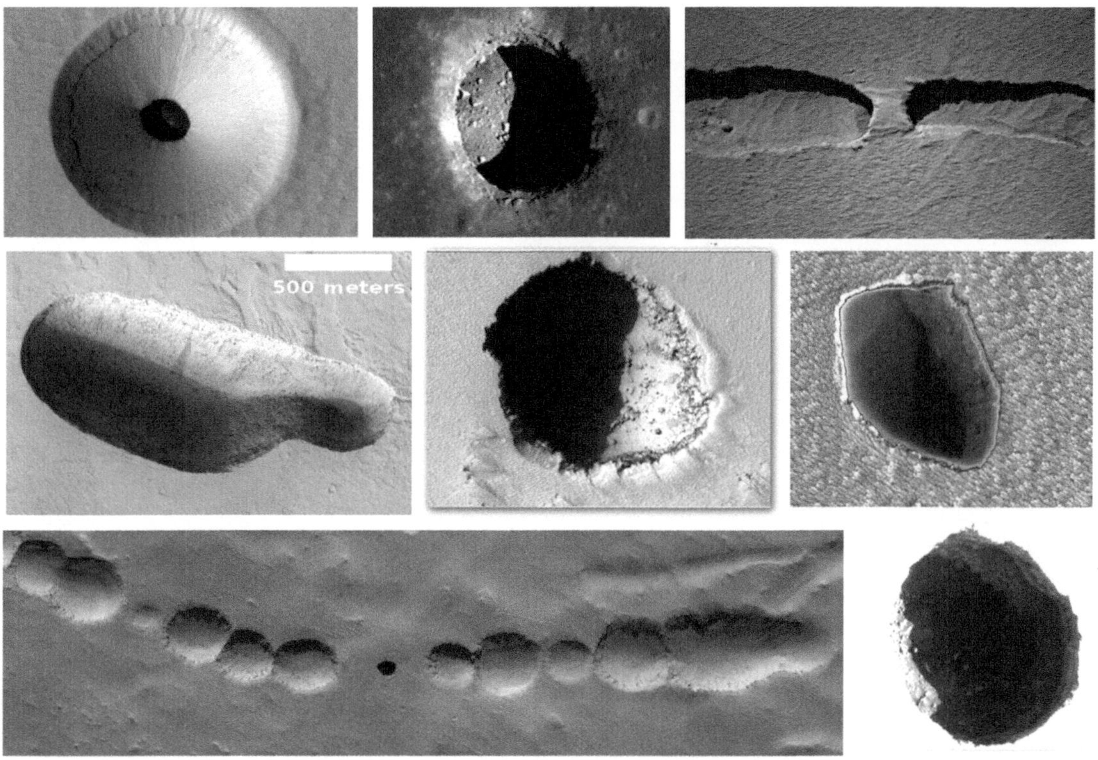

FIGURE 10.1 Sinkholes on Mars (compiled from images released by NASA).

10.1.2 Characteristics of Volcanic Rocks on Mars and the Moon

Mars explorer rovers (MERs, Opportunity, Spirit, Curiosity and Perseverance) have provided some scientific data on the composition and characteristics of rocks in the surveyed areas of Mars. Figure 10.2 shows views of basaltic rocks. As noted from the figure, basalts of Mars are quite porous, which may be vesicular basalt. Similarly, images of rocks from the Moon revealed the existence of vesicular basalts, as shown in Figure 4.6.

It is very likely that the porous structure may lead to heavy weathering as well as a decrease of mechanical properties. As there are no experimental data on basalts from Mars, the drilling and grinding operations by the rovers on the rocks of Mars may be utilized to infer their mechanical properties, as discussed in Chapter 13. Figure 10.3 shows views of vesicular basalt samples from different parts of Earth. Figure 10.4 shows the variation of uniaxial compression strength (UCS) and Brazilian tensile strength of basalt sampled from Japan and Türkiye as a function of porosity. As noted from Figure 10.3, the basalts of Mars and the Moon resemble those of Earth. Therefore, the mechanical properties of the vesicular basalt of Earth may be utilized to evaluate the mechanical properties of the vesicular basalts of Mars and the Moon.

FIGURE 10.2 Views of Martian basalts.

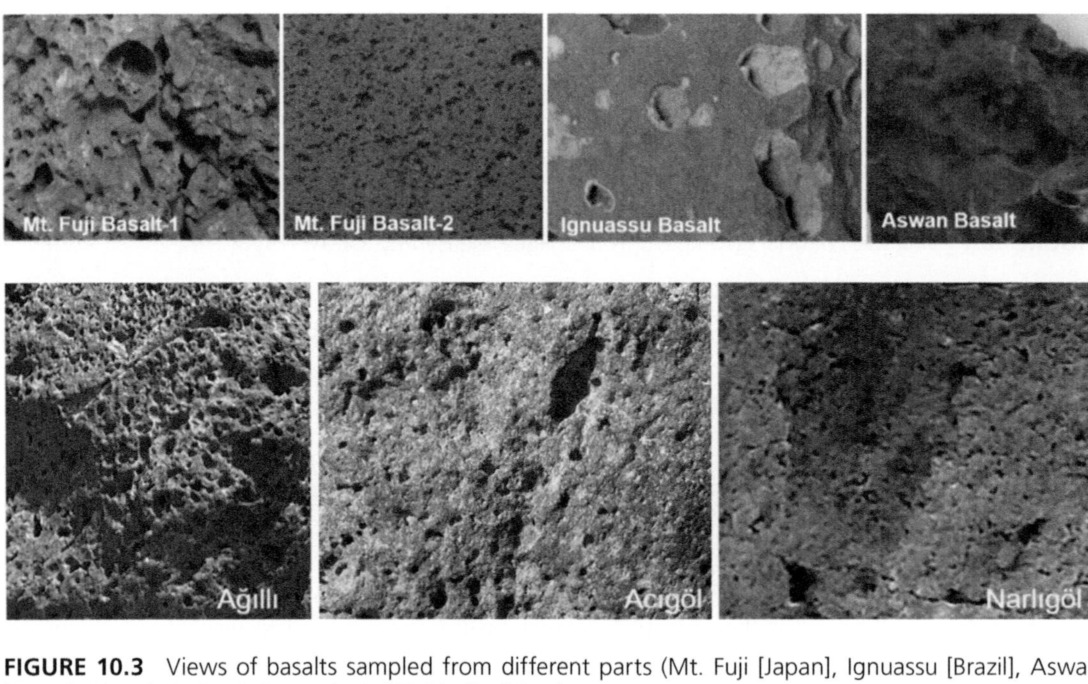

FIGURE 10.3 Views of basalts sampled from different parts (Mt. Fuji [Japan], Ignuassu [Brazil], Aswan [Egypt], Acıgöl, Narlıgöl, Cappadocia [Türkiye]) of Earth.

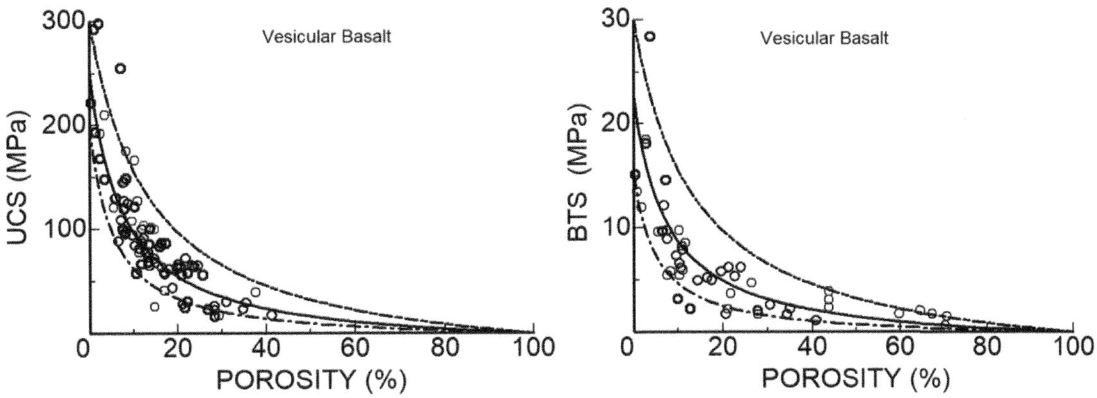

FIGURE 10.4 Variation of (a) uniaxial compressive strength and (b) Brazilian tensile strength of basalt from Türkiye and Japan with porosity.

10.2 ROCK MASS CONDITIONS AND THEIR PROPERTIES

As discussed earlier by Aydan (2017c, 2020a, 2021a, 2021b, 2022a, 2022b), rocks, discontinuities and rock masses of Mars and the Moon are quite similar to those of Earth. Therefore, it should be possible to evaluate their mechanical properties utilizing the methods developed for those of Earth (Barton et al. 1974; Bieniawski 1973, 1989). Aydan et al. (2014b) recently developed a new rock classification and proposed a procedure for evaluating mechanical properties (ψ_m) of rock masses in terms of intact rock properties (ψ_i) and a state variable-like parameter for rock masses, which is called the new rock mass quality rating (RMQR) using the following formula:

$$\frac{\psi_m}{\psi_i} = \alpha_{100} - (\alpha_{100} - \alpha_0)\frac{RMQR}{RMQR + \beta(100 - RMQR)} \tag{10.1}$$

where α_0 and α_{100} normalized properties for the values of RMQR as 0 and 100. β is an empirical coefficient. Their values change according to the property of rock mass under consideration. On the bases of in situ tests on rock masses, Table 10.1 summarizes the values of parameters for each respective property of rock masses. Horiuchi et al. (2019) advanced Eq. (10.1) by taking into account the porosity of intact rock:

$$\frac{\psi_m}{\psi_i} = \alpha_{100} - (\alpha_{100} - \alpha_0)\frac{RMQR}{RMQR + \beta(100 - RMQR)} \cdot \left(1 - \frac{n}{n + \theta(100 - n)}\right) \tag{10.2}$$

where n and θ are the porosity and empirical coefficient. The value of coefficient θ is generally 1 or greater depending upon the rock type and intact rock property.

TABLE 10.1 Values of α_0,α_{100} and β for Various Properties of Rock Mass

PROPERTY (α)	α_0	α_{100}	β
Deformation modulus	0.0	1.0	6
Poisson's ratio	2.5	1.0	0.3
Uniaxial compressive strength	0.0	1.0	6
Tensile strength	0.0	1.0	6
Cohesion	0.0	1.0	6
Friction angle	0.3	1.0	1.0

10.3 STABILITY ASSESSMENT TECHNIQUES FOR UNDERGROUND OPENINGS

In this section, some stability assessment techniques are presented. Specifically, empirical and analytical methods are described. In addition, numerical methods such as the finite element method, discrete finite element method and other discrete numerical methods are available. However, although they are not presented here, interested readers can consult certain publications (e.g., Kawamoto and Aydan 1999; Kawamoto et al. 1991; Aydan et al. 1996a; Aydan and Tokashiki 2011; Lang 1994).

10.3.1 Empirical Methods Based on Rock Classifications

The empirical techniques generally involve rock mass classifications. Several empirical relations between allowable span or height and rock mass quality value (Q) or rock mass rate (RMR) have been proposed in the literature. Barton et al. (1980) suggested an empirical formula for the non-supported span (L) of the underground openings in relation to the Q-value:

$$L = 2 \cdot ESR \cdot Q^{0.4} \tag{10.3}$$

where ESR is the equivalent support ratio. When the value of ESR is 1, it corresponds to the empirical line for the unsupported span limit in the original article of the Q-system by Barton et al. (1974). Barton (1976) also suggested the following formula between the unsupported span (L) (unit is m) and Q-value on the basis of his observations on the Carlsbad karstic caves in USA:

$$L = 66 \log Q + 2 \tag{10.4}$$

Lang (1994) proposed two empirical bounds for the span between stable and unstable openings on the basis of observations in mines using the RMR. However, he did not provide any specific equation for his bounds. When these empirical relations are applied to natural rock underground structures, they would generally yield very conservative results. Furthermore, the physical interpretation of the empirical bounds is very difficult in practice. One may find numerous examples such that the natural rock underground structures are stable although the data of caves may require some levels of rock support.

Karstic caves may present some engineering problems especially in urbanized areas along shorelines and riversides. The effect of karstic caves on the overall stability of the castle remains were investigated using empirical, analytical and static and dynamic numerical techniques by Aydan and Tokashiki (2011). In addition, one-dimensional shaking table experiments on model caves were performed, and the effect of cavity filling as a countermeasure was investigated. These investigations clearly demonstrated that the filling of cavities should be an effective measure to achieve the overall safety of superstructures.

On the basis of observations on the Karstic natural caves in Ryukyu limestone, Tokashiki (2011) and Aydan and Tokashiki (2011) defined four different stability categories for natural underground openings in Ryukyu limestone utilizing the RMR rock classification system (Bieniawski 1973, 1989). However, Aydan (2018b) updated this system and related it to the RMQR and increased the stability category to 6, as shown in Figure 10.5. The limiting lines of each category are fitted to the following two formulas for the span (L) (unit is m) of natural underground openings as a function of RMQR:

Linear Function Formula: $L = a\text{RMQR} + b$ (10.5)

Power Function Formula: $L = a\text{RMQR}^b$ (10.6)

where a and b are empirical coefficients. The values of the parameters are given in Tables 10.2 and 10.3 for the caves in Ryukyu limestone (UCS = 25 MPa) and the Alpine limestone (UCS = 60 MPa). When these empirical relations are used for sinkholes (skylights) on Mars and the Moon, the properties of the surrounding rock mass are utilized. When uniaxial compressive strength and tensile strength are selected, the porosity of vesicular basalt should be taken into account. The porosity of basalts can be determined from image analyses of those shown in Figures 4.6 and 10.2. The relations shown in Figure 10.4 may be utilized to determine the values of strength for stability assessment computations, which are given in the next subsections.

TABLE 10.2 Values of Empirical Constants

	$L = a\text{RMQR} + b$		$L = a\text{RMQR}^b$	
LINES	*a*	*b*	*a*	*b*
I–II	0.58	−50.0	0.008	1.5
II–III	0.60	−42.0	0.018	1.5
III–IV	0.62	−32.0	0.030	1.5
IV–V	0.64	−22.0	0.042	1.5
V–VI	0.66	−6.0	0.060	1.5

TABLE 10.3 Values of Empirical Constants in Eqs. (10.5) and (10.6)

	LINEAR FUNCTION		*POWER FUNCTION*	
LINES	*a*	*b*	*a*	*b*
I–II	2.0	−120	0.007	2.0
II–III	2.4	−120	0.012	2.0
III–IV	2.8	−112	0.017	2.0
IV–V	3.2	−96	0.022	2.0
V–VI	3.6	−72	0.029	2.0

Category	State	$\frac{H_f}{H_r}$	$\frac{B_s}{B}$	Comments
I		0.0	0.0	Opening locally and globally stable
II		0.0-0.1	0.0	Some rock block falls from roof. Opening globally stable
III		0.1-0.3	0.0	Block falls from roof and sidewalls into the opening occur and the failure zone increases in size. Roof height is higher than opening width. The failure zone may increase in size with time.
IV		0.3-0.7	0.0	Considerable scale of falls and sliding of rock blocks from the roof and sidewall of openings occur and the failure zone larger in size. Roof height is much higher than opening width. The failure zone may increase reach ground surface in long-term
V	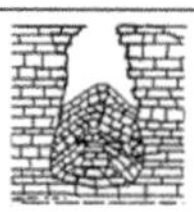	0.7-1.0	0.8-1.0	Failure zone reaches to ground surface and a small size crater develops at ground surface. The possibility of collapse zone may increase in size and shoulder may fall into opening in long-term
VI		≥1.0	>1.0	Opening globally unstable. In other words, it is in a total collapse state. Deep Sinkhole appear on the ground surface.

(a) (b)

FIGURE 10.5 (a) Categorization and (b) comparison of empirical bounds for the state of karstic caves.

10.3.2 Analytical Methods

The stability assessment methods for the roofs of shallow underground openings are generally based on the bending theory of beams or arching theory used in structural mechanics (Figure 10.6). Particularly, the arching theory has been popular in mining engineering and many formulations are developed with the consideration of various modes of failure (i.e., Aydan 1989, 2008; Kawamoto et al. 1991; Aydan and Tokashiki 2007, 2011).

The limit of roof span (L) normalized by roof rock layer thickness (h_r) under its dead weight and topsoil with thickness of (h_s) can be obtained in the following form:

$$\frac{L}{h_r} = \sqrt{\lambda \frac{\sigma_t}{(\gamma_r h_r + \gamma_s h_s)}} \tag{10.7}$$

where σ_t is the tensile strength of roof layer. The value of λ would take the following values 1/3, 2/3 and 2 for cantilever, simple and built-in beams. The cantilever beam situation may arise when the roof of the opening is crossed by a fault, which is a common situation in karstic caves.

The arching theory for assessing the stability of roofs of shallow underground openings is generally based on three-hinged beams. As discussed by Aydan (1989, 2008), and Aydan and Tokashiki

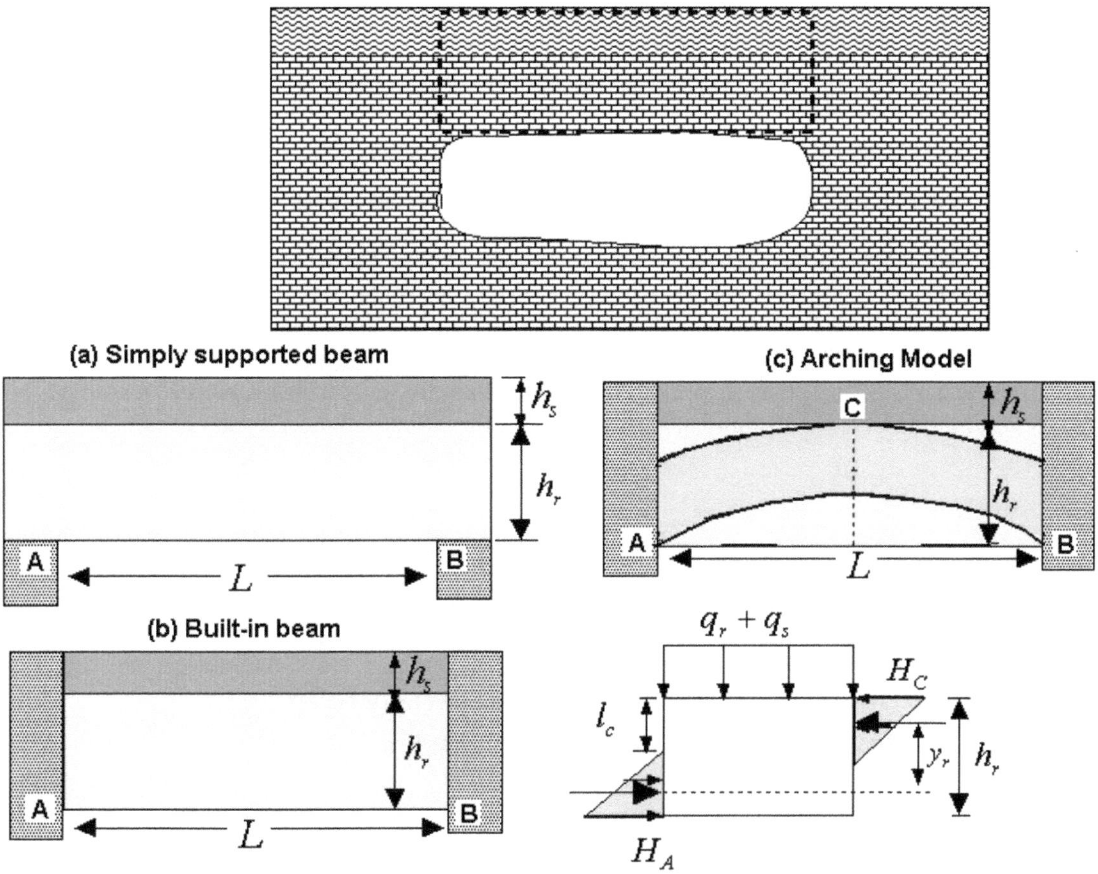

FIGURE 10.6 Illustrations of mechanical models for theories of the bending and arching of beams.

(2011), the rotation of rock layer, in which arch is formed, needs to be initiated for the arch action to take place. The failure modes of an arch are various; they are crushing at the crown and/or abutments, vertical or horizontal sliding at abutments and sliding along an existent discontinuity within the arch. The detailed derivations for each failure mode are presented in the doctorate thesis of Aydan (1989) and in the article by Kawamoto et al. (1991). The classical arching theory assumes that the failure mode of the arch takes place by crushing at the crown or abutments. The final form for the limit of roof span (L) normalized by rock roof layer thickness (h_r) under its own weight and topsoil with thickness of (h_s) is as follows:

$$\frac{L}{h_r} = \sqrt{\lambda \frac{\sigma_c}{(\gamma_r h_r + \gamma_s h_s)}} \tag{10.8}$$

where σ_c is the compressive strength of roof layer. The value of λ would take the values 4/3 when the crack length (ℓ_c) is zero. However, the maximum resistance of the arch is attained when the value of λ is 3/2 after some manipulations of the crack length in the arch. The effect of concentrated loads can be taken into account if necessary.

If Terzaghi's arching theory is adopted for cohesive (c) and frictional (ϕ) ground conditions, one can also derive the following relation (Terzaghi 1943).

$$\frac{L}{h_r} = 2\left(\frac{c}{\gamma} + K_h \tan\phi\right) \tag{10.9}$$

Theoretically, the lateral stress coefficient (K_h) would be equal to 1. The tensile, uniaxial compressive and cohesive strengths and friction angle of the rock mass in Eqs. (10.5)–(10.7) can be obtained using the formulas suggested by Aydan et al. (2014b, 2017) in terms of those of intact rock and the RMQR value of rock masses. It should be noted that the bending and arching theory should not be applied when the value of RMQR is less than 30 due to the poor conditions of rock masses (Aydan et al. 2017). Furthermore, the bending theory provides a theoretical basis for local stability, while the arching theory theoretically establishes the limit spans for the collapse of natural underground openings.

10.4 APPLICATIONS TO SINKHOLES ON EARTH, MARS AND THE MOON

10.4.1 Comparison of Empirical and Analytical Estimations for Sinkholes in Earth

First, the sinkhole issue on Earth is analyzed. The case history data were obtained from sinkholes observed in the vicinity of Mt. Fuji in Japan, Mt. Ağrı, Karacadağ and Hatay region in Türkiye, Hawaii and mainland US (Figure 10.7). The author personally visited some of sinkholes and volcanic caves in Japan, Türkiye and Hawaii. In addition, some data were gathered from Azerbaijan, Taiwan, Iceland.

FIGURE 10.7 Locations and views of some sinkholes in the vicinity of Mt. Fuji, Japan, in the vicinity of Mt. Fuji in Japan, Mt. Ağrı in Türkiye, and mainland US.

The geometry and size of sinkholes are evaluated, and their stability is categorized according to Figure 10.5. For sinkholes and caves on Earth, RMQR values are determined. Figure 10.8 compares observational data on caves and sinkholes on Earth with empirical bounds given by Eqs. (10.3)–(10.4), together with constants given in Table 10.2. Both linear and power functions provide good estimations for observational data.

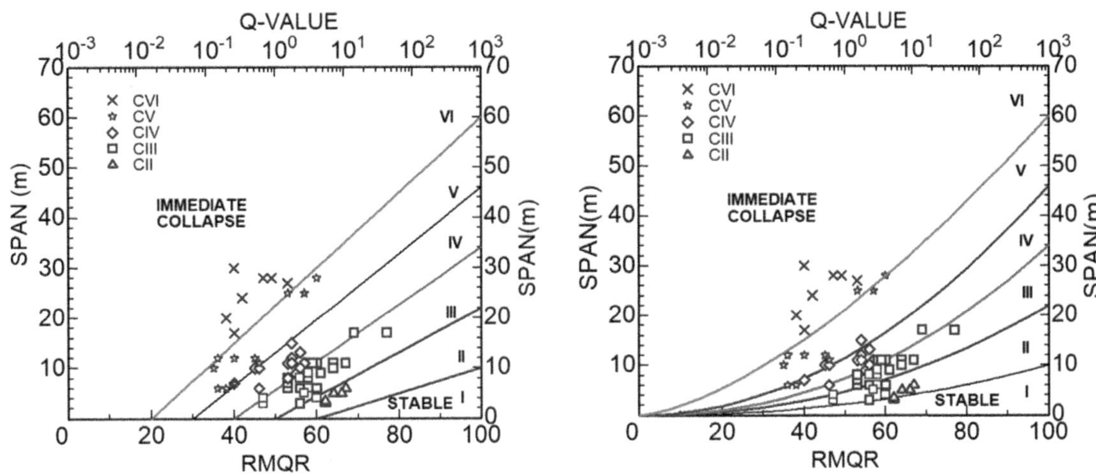

(a) Linear function

(b) Power function

FIGURE 10.8 Comparison of empirical stability charts with observation data on Earth.

The stability estimations are based on analytical methods compared with observational data for sinkholes in Figure 10.9 for the given parameters for intact rock and rock mass. It seems that the arching theory provides an upper bound for the formation of sinkholes provided that the material properties of intact rock and rock mass are properly evaluated. Nevertheless, more detailed explorations and analyses of each site should provide better estimations.

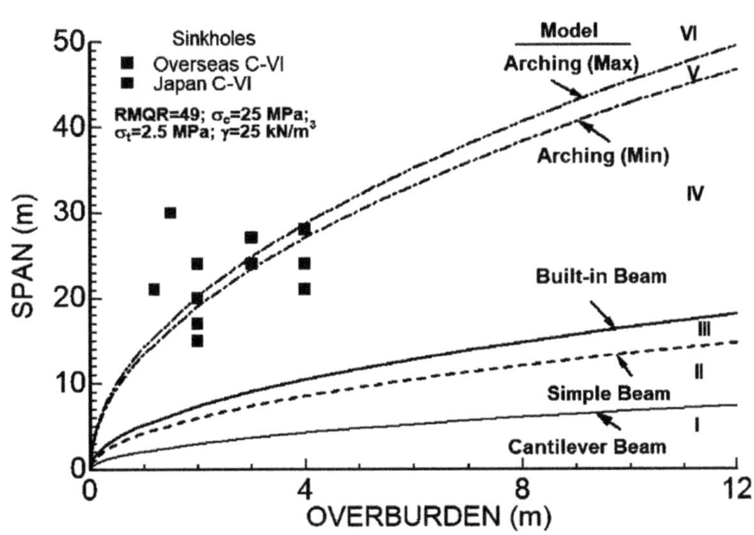

FIGURE 10.9 Comparison of analytical stability charts with the observation data of sinkholes on Earth.

10.4.2 Comparison of Empirical and Analytical Estimations for Sinkholes in Mars and Moon

Cushing (2012) and Cushing et al. (2007, 2015) investigated the sinkholes of Mars and provided a detailed information about each sinkhole. Similarly, Cruikshank and Wood (1971), Wagner and Robinson (2014) provided information on sinkholes, which they call lunar pits. However, some additional data are necessary for stability analyses. For such a purpose, important parameters are the thickness of the cover rock and the rock mass state variable. Some attempts have been done to evaluate these two parameters from images on the bases of knowledge available of sinkholes of Earth. Furthermore, the strength of vesicular basalt is estimated to be about 25 MPa in view of its porosity and the experiments on Mt. Fuji basalt. As the gravity coefficient of Mars and the Moon is 0.387 and 1/6 times that of Earth, the stress induced by the Mars and the Moon gravity is assumed to be 0.387 and 1/6 times that on Earth, respectively. By considering these facts, the empirical and analytical methods have been used to assess the sinkholes occurrence for Mars. The estimations from empirical and analytical methods are shown and compared with observational data in Figure 10.10. The overall trend is quite promising despite the fact that many data are based on some inferences for similar situations on Earth. Once again it is noted that the analytical methods yield also the bounds to be on the safe side. When observational data are compared with empirical estimations, the overall trends are quite similar, and the empirical estimation is on the safe side.

Despite the fact that the rock mass classification for the collapsed caves in the Moon could not be done due to limited observations, some comparisons with empirical relations were done, as shown in Figure 10.11a. Computations were done for analytical methods by assuming that gravitational acceleration is 1/6 g and that the RMQR value is 60, as shown in Figure 10.11b. The estimations from analytical methods for given materials properties shown in the figure show similar trends with observational data. As noted, analytical methods also yield the bounds on the safe side.

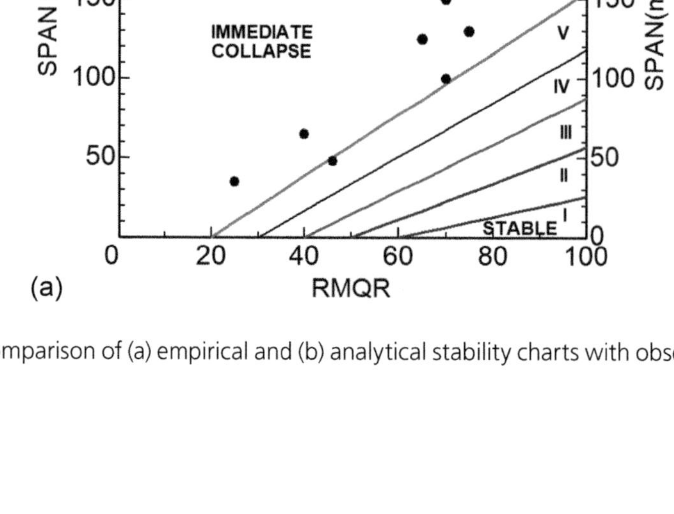

FIGURE 10.10 Comparison of (a) empirical and (b) analytical stability charts with observation data on Mars.

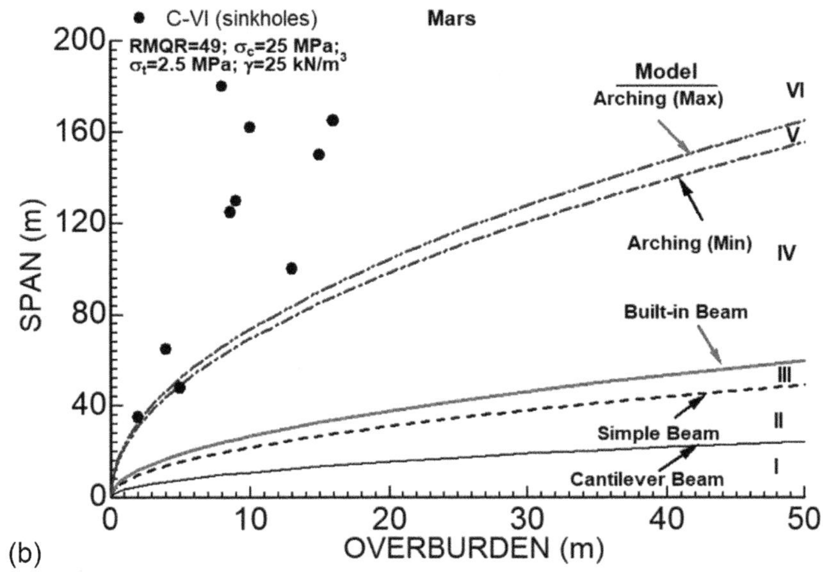

FIGURE 10.10 (Continued)

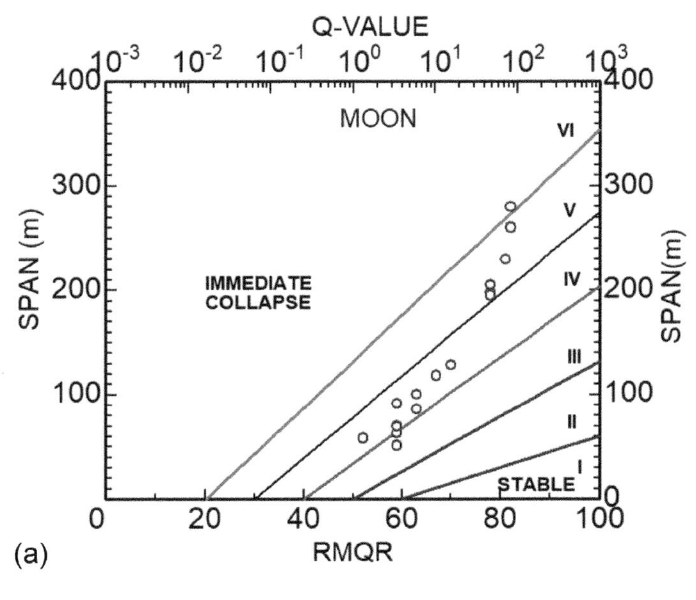

FIGURE 10.11 Comparison of (a) empirical and (b) analytical stability charts with observation data on the Moon.

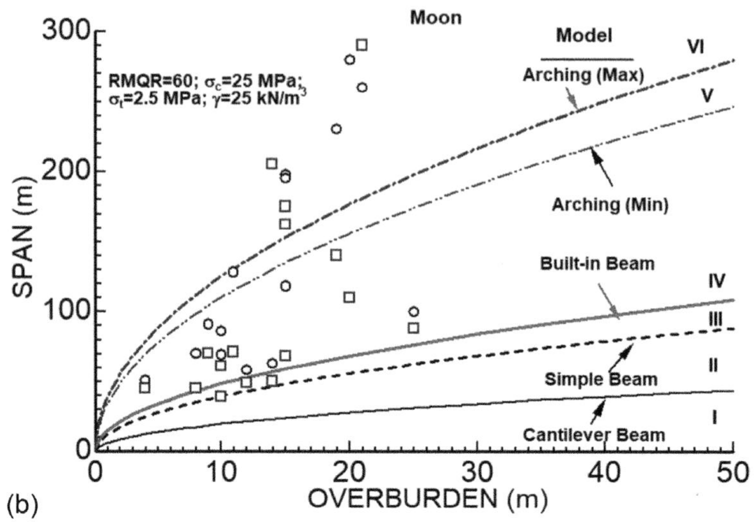

FIGURE 10.11 (Continued)

10.5 NUMERICAL METHODS

The stability of underground openings can be analyzed using some available techniques. The most commonly used techniques are finite element methods, which may be the conventional continuum type or the discontinuum type, in which special elements such as joint, interface or contact elements are used. Figures 10.12 and 10.13 show some applications of these methods to shallow underground caves at the Kuşini antique marble Quarry (Selçuk, Türkiye) and Kaklık Karstic Cave (Denizli, Türkiye) on Earth (Aydan and Kumsar 2005; Kumsar and Aydan 2022).

10.6 SINKHOLES (SKYLIGHTS) IN VENUS

As mentioned previously, the surface topography of Venus has been obtained through radar imaging by Magellan spacecraft. Figure 10.14 shows several images with possible collapsed volcanic channels, which may be interpreted as sinkholes or skylights on Venus. Although this interpretation may be speculative, there is no doubt that such structures must exist on Venus in view of similar observations on other planetary bodies.

10.7 SINKHOLES MERCURY

Despite adverse conditions for imaging Mercury due to the close proximity of the planet to the Sun, imposing high temperature and high solar pull on the spacecraft, Mercury was surveyed by Messenger

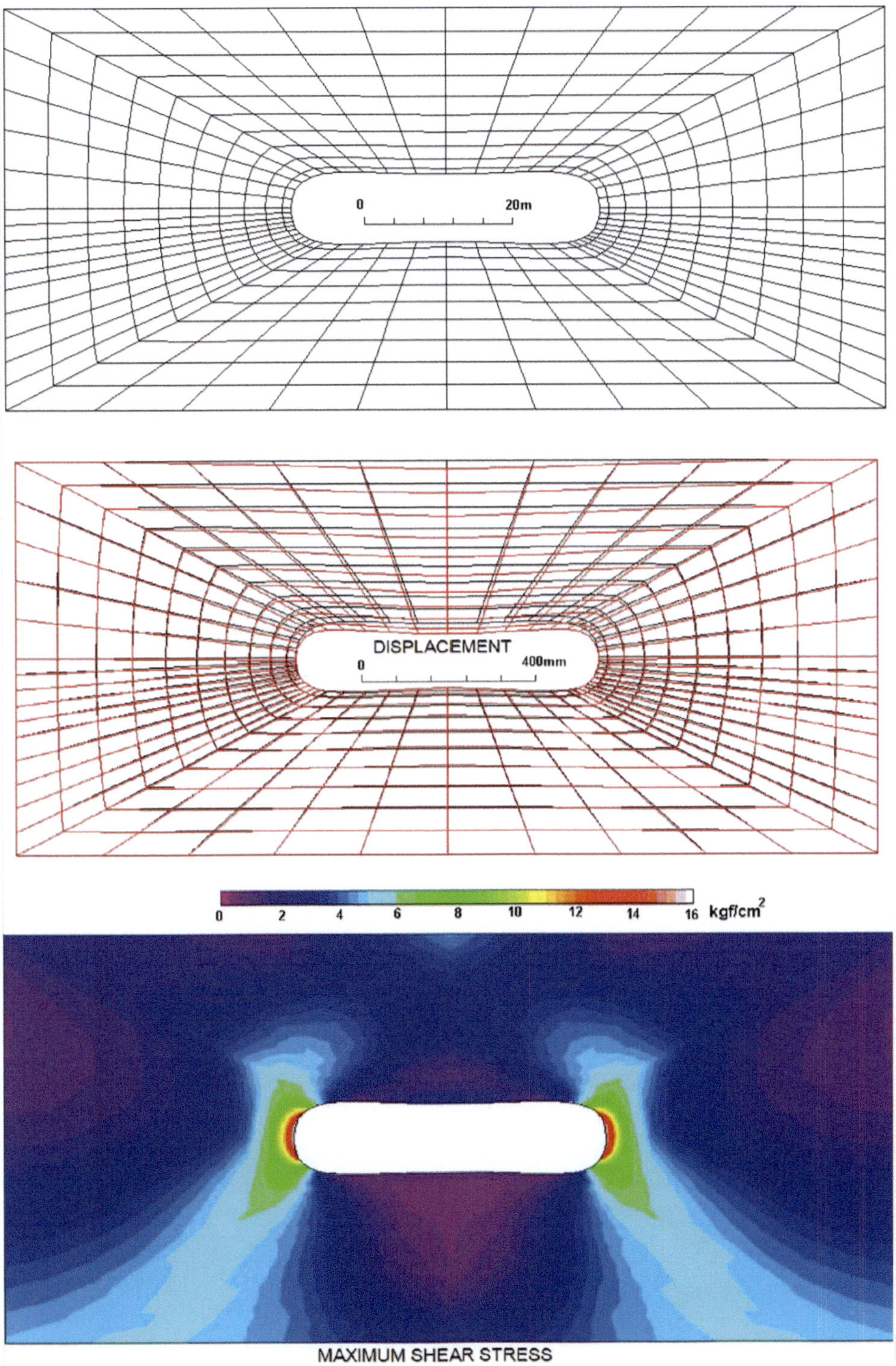

FIGURE 10.12 Mesh, deformed configuration and maximum shear stress distribution of Kuşini antique marble Quarry by the finite element method (Aydan and Kumsar 2005).

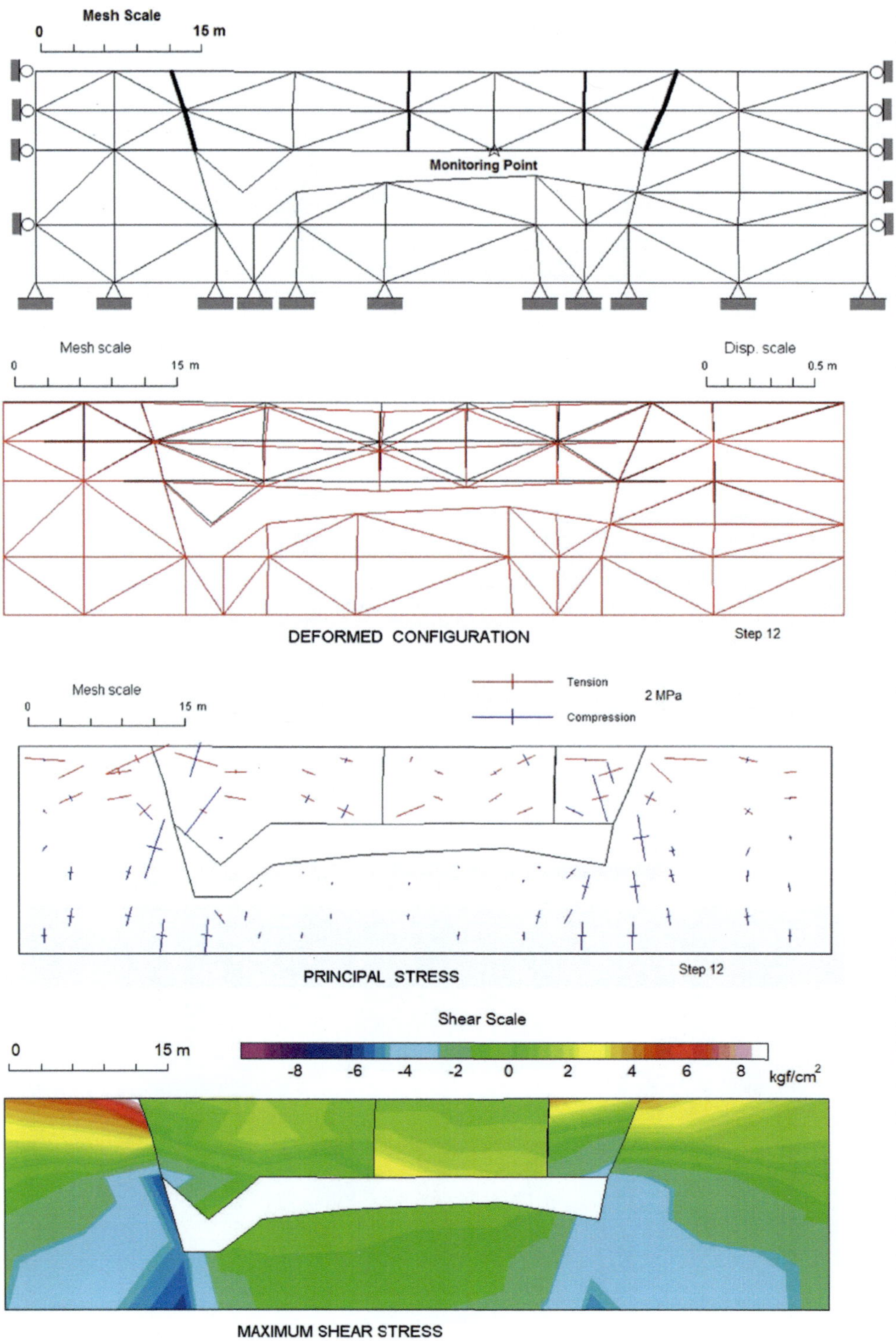

FIGURE 10.13 Mesh, computed deformed configuration, principal stress and maximum shear stresses at Step 12 for Kaklık Karstic Cave by the discrete finite element method (DFEM) (from Kumsar and Aydan 2022).

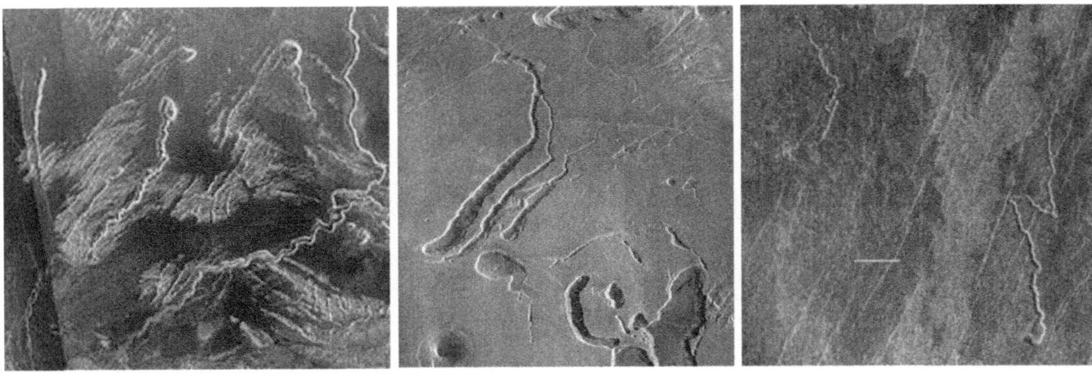

FIGURE 10.14 Collapsed lava channels in Venus.

spacecraft. Figure 10.15 shows several images with possible collapsed volcanic channels, which may be interpreted as sinkholes or skylights on Mercury. This interpretation may seem to be speculative. Nevertheless, there should be no doubt that such structures must exist in Mercury in view of similar observations on other planetary bodies.

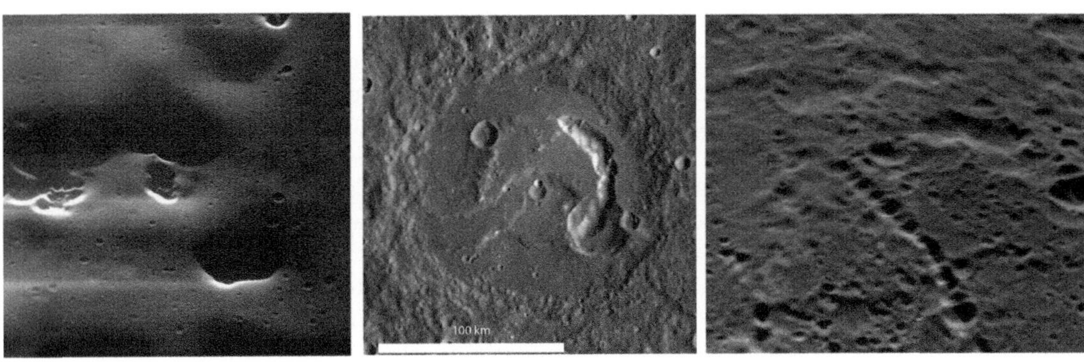

FIGURE 10.15 Collapsed lava channels in Mercury.

Characteristics of Hydro-Magnesite of Salda Lake (Türkiye) and Their Implications on Rocks and Jezero Crater in Mars

11

This chapter is based on the outcomes of findings by the author and his colleagues (Kumsar et al. 2023) on a hydro-magnesite formation in Salda Lake in Türkiye, which receive great attention as an example on Earth in relation to the early life traces in the Jezero Crater of Mars. The authors investigated Salda Lake and sampled hydro-magnesite. The sampled stromatolites vary from newly formed soft ones to quite hardened ones. Microscopic structures and X-ray analyses of samples were used to characterize the mineralogical constituents. Some samples were prepared from large pieces of stromatolites to determine physico-mechanical properties.

Thin sections were prepared for evaluating microscopic structures and X-ray analyses of samples were used to characterize the mineralogical constituents. Some samples were prepared from large pieces of stromatolites to determine the physical properties, such as density, porosity, elastic wave velocity, and mechanical properties such as Brazilian tensile strength (BTS), three-point bending tensile strength, uniaxial compression strength (UCS) and elastic modulus. Experimental results indicated that the mechanical properties strongly depend upon the porosity of samples.

The investigations into the various characteristics of hydro-magnesite obtained from the stromatolites of Salda Lake are presented, and various interrelations as well as the yield strength envelope of hydro-magnesite are evaluated. The implications of the outcomes from these investigations on the characteristics of rocks in Jezerro Crater in Mars are discussed. This study may be the first rock mechanics study on hydro-magnesites.

11.1 GEOGRAPHY AND GEOLOGY OF SALDA LAKE

Salda Lake is located within Burdur Province of Türkiye, which is also known as Lake District. It is a closed system, and three creeks flow into the lake (Figure 11.1). The lake covers an area of 4370 hectares, and its depth reaches 196 m. The present level of the lake is approximately 10–12 m lower as inferred from the deposition of hydro-magnesites along the shores in the Kayadibi location.

DOI: 10.1201/9781003643852-11

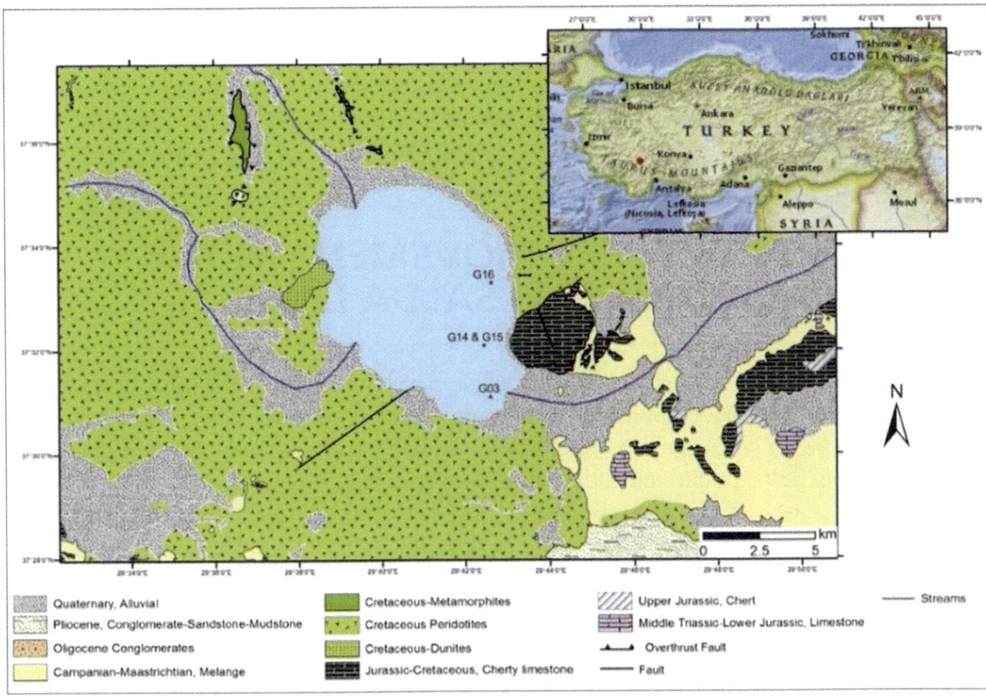

(a) Location and simplified geology (from Erentöz 1974)

(b) Kayadibi

FIGURE 11.1 (a) Location and simplified geology of the vicinity of Salda Lake and (b) a view of the Kayadibi site.

The base of the lake consists of serpentinized ultramafic terrain consisting of mainly harzburgite. The area is tectonically and hydrothermally very active. Hydro-magnesite deposits are widely observed at the Kocaadalar and Kayadibi localities with widespread stromatolites occurrences, while the deposits cover the entire perimeter of the lake. The stromatolites are porous, and their surfaces are cauliflower-like.

11.2 TECTONIC MOVEMENT IN THE VICINITY OF SALDA LAKE

As noted from Figure 11.1(b), at least four erosion levels can be recognized. These erosion level changes can be related to water level fluctuations due to climatic changes, tectonic movements or both. During site investigations, one may easily note slicken-sided features in serpentinized ultramafic rocks at Doğanbaba, Salda and Sultanpınarı, indicating uplift movements during the closure of the Thetis Ocean in the geologic past. One can also observe strike-slip faults in limestone as well as hydro-magnesite deposits along the roadway at the Kayadibi locality (Figure 11.2). The strike-slip fault is noted in satellite images as well as in Figure 11.1(a). The dip direction and dip angle of the strike slip fault range between 210–230 and 70–85 degrees, respectively, while the rake angle of striations ranges between 5 and 30 degrees. The fault is inferred to be strike-slip with a normal component, and it is probably the youngest tectonic feature in the vicinity of Salda Lake. As a result of the tectonic movements, the hydro-magnesite deposits at higher elevations are expected to be older than those along the shoreline of Salda Lake. It is also expected that the hydro-magnesite deposits would harden due to water content loss.

FIGURE 11.2 Striations in hydro-magnesite and limestone along the roadway at Kayadibi locality.

11.3 SAMPLING AND GEOCHEMICAL AND THIN SECTIONING ANALYSES

Sampling was done mainly at Kayadibi locality on hydro-magnesite stromatolites and deposits, as shown in Figure 11.3. Sampling was directly done from stromatolites that were underwater,

(a) (b) (c)

FIGURE 11.3 Views of sampling locations (a) underwater and (b, c) above the water level.

below the lake level, and those above the lake level for geochemical and thin sectioning analyses for the microscopic examination of composition and structure of formed hydro-magnesites. Furthermore, large rock hydro-magnesite blocks were gathered for physico-mechanical property testing samples. The samples were brought to Japan for geomechanical and thin sectioning analyses, physico-mechanical tests, and further analyses and investigations. Geochemical analyses indicated that the samples consist of carbonates containing magnesium and calcium with a small amount of silica.

Figure 11.4 shows microscopic views of samples from stromatolites under and above water, whereas Figure 11.5 shows the samples and thin sections of samples used in the physico-mechanical tests. Diatoms and the needle-like crystal structure of hydro-magnesites were noted and, further, the very porous structure of formed hydro-magnesites. Hydro-magnesites are quite soft when they are underwater.

(a) Views of samples underwater (b) Views of samples above the water level

FIGURE 11.4 Macroscopic and microscopic views of samples underwater and above the water level.

11.4 PHYSICO-MECHANICAL TESTS

Experiments were carried out on samples prepared for Brazilian tensile, uniaxial compression, three-point bending tests according to the suggested aspect ratios of ISRM Suggested Methods (ISRM 2007), even though the sample sizes could not be in accordance with ISRM Suggested Methods due to limitations for sampling. The unit weight of samples was determined, and the results are given in Table 11.1. Figure 11.6 shows examples for Brazilian, three-point bending and uniaxial compression experiments. In each testing group, samples exhibited stick-slip type behavior due to partial pore collapses.

TABLE 11.1 Experimental Results for Unit Weight of Samples in Uniaxial Compression, Brazilian and Three-point Bending Tests

SAMPLE TYPE	*LOWEST*	*HIGHEST*	*MEAN*
Uniaxial compression (UCS)	11.0	12.8	11.68
Brazilian (BTS)	7.5	11.3	9.88
Three-point bending	9.54	11.6	10.8

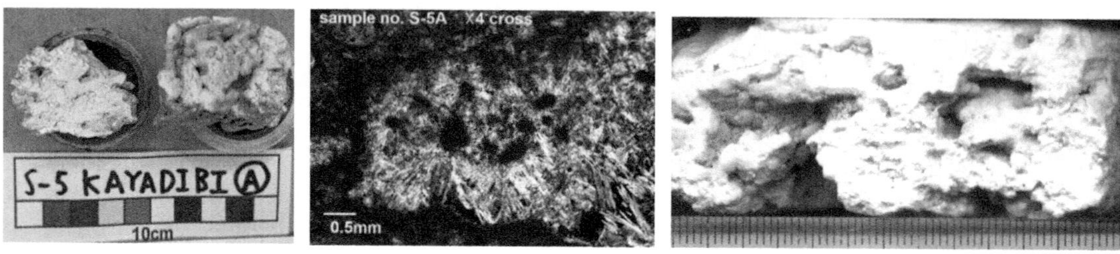

(a) (b) (c)

FIGURE 11.5 (a) Macroscopic and (b) microscopic views of Sample S-5 and (c) a macroscopic view of a sample for physico-mechanical property tests.

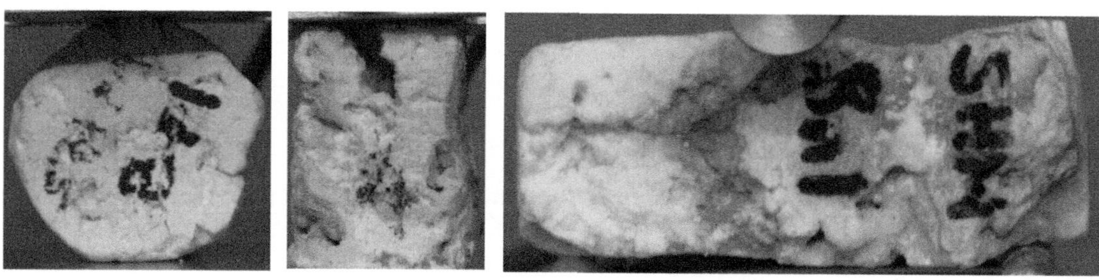

(a) Brazilian (b) Uniaxial compression (c) Three-point bending

FIGURE 11.6 Views of samples during mechanical tests.

11.4.1 Porosity

If the Salda Lake deposits is assumed to be purely hydro-magnesite, the likely porosity of samples may be computed from the following relation, provided that the unit weight of hydro-magnesite minerals is known:

$$n = \left(1 - \frac{\gamma_b}{\gamma_g}\right) \times 100 \quad (11.1)$$

where γ_b, γ_g are bulk unit weight and grain unit weight, respectively. The unit weight of hydro-magnesite ranges between 21.6 and 22.0 kN/m^3 with a mean value of 21.8 kN/m^3. Thus the porosity of Salda Lake hydro-magnesite deposits is likely to range between 41.3 and 65.6%.

11.4.2 Brazilian Tensile Tests

A number of Brazilian tests were carried out on Salda Lake hydro-magnesite samples. Figure 11.7 shows the displacement versus tensile stress responses. As noted from the figure, Brazilian tensile strength (BTS) values range between 281.23 and 573.82 kPa with a mean of 437.08 kPa. As noted from these values, the hydro-magnesite samples are quite weak in terms of mechanical properties.

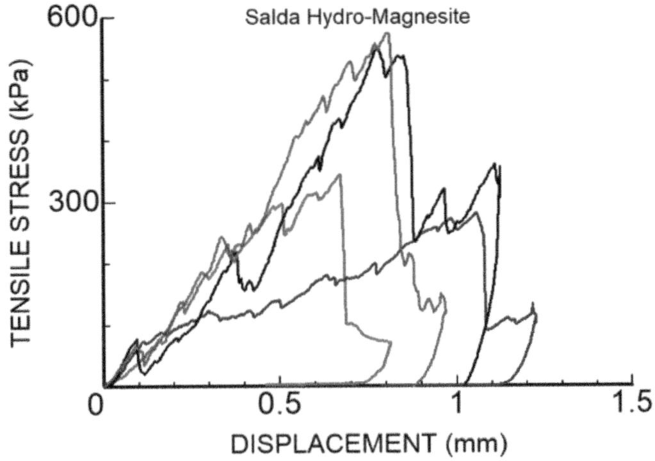

FIGURE 11.7 Displacement–stress responses in Brazilian tests.

11.4.3 Uniaxial Compression Tests

A number of uniaxial compressive strength experiments were carried out, and the strain–stress responses measured during experiments are shown in Figure 11.8. As noted from the figure, the uniaxial

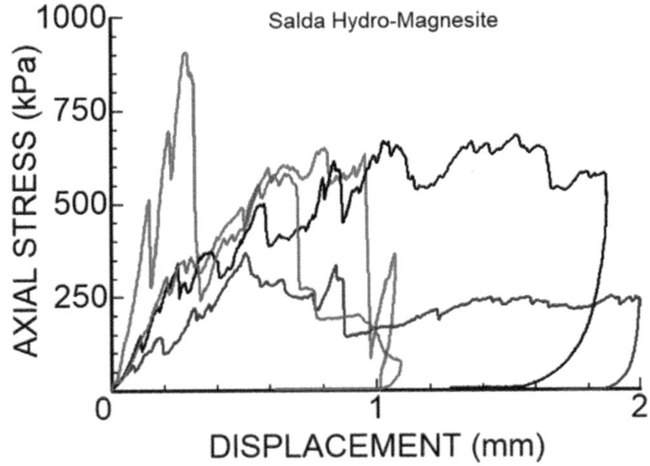

FIGURE 11.8 Strain–stress responses in uniaxial compression tests.

compressive strength (UCS) values of Salda Lake hydro-magnesite samples range between 371.0 and 906.4 kPa with a mean of 635.5 kPa. Similarly, the uniaxial compressive strength of hydro-magnesite samples is also quite weak. Nevertheless, the samples observed at higher elevations seem to have greater strength through some physical inspections. However, this aspect needs to be checked through mechanical tests.

11.4.4 Three-point Bending Tests

Three three-point bending tests were carried out, and the displacement–stress responses were measured during experiments, as shown in Figure 11.9. As noted from the figure, bending strength values of Salda Lake hydro-magnesite samples range between 415.0 and 657.9 kPa with a mean of 566.5 kPa. Tensile strength obtained from three-point bending tests are always higher than that obtained from Brazilian tensile tests. This observation is in accordance with the previous experimental results by the authors (Aydan 2022c).

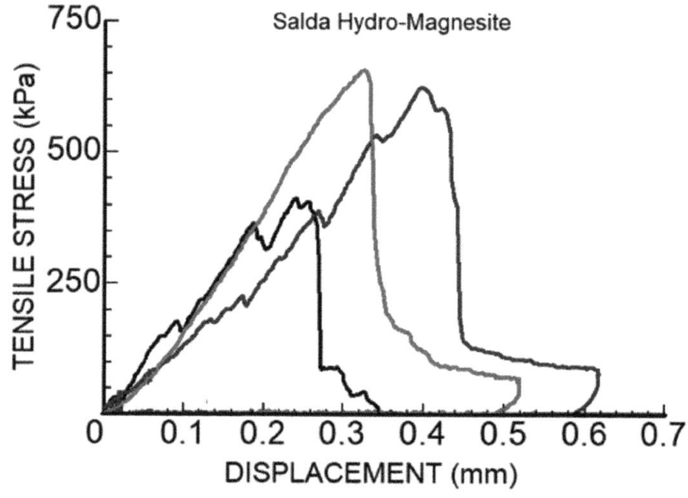

FIGURE 11.9 Displacement–stress responses in three-point bending tests.

11.4.5 Mohr–Coulomb Strength Envelopes

The Mohr–Coulomb failure criterion (e.g., Aydan 2020a) was selected, and the values of the Mohr–Coulomb failure criterion for Salda Lake hydro-magnesite was evaluated by utilizing lower, mean and peak Brazilian tensile strength (BTS) and uniaxial compression strength (UCS) values. The results are shown in Figure 11.10. The cohesion for lower, mean and peak values are 0.16, 0.26 and 0.36 MPa, and these values are quite similar to the tensile strength values. Similarly, the friction angles for lower, mean and peak values of BTS and UCS are 7.9, 10.7 and 13.0 degrees. Although other failure criteria may be utilized, Mohr–Coulomb failure criterion is sufficient for engineering purposes.

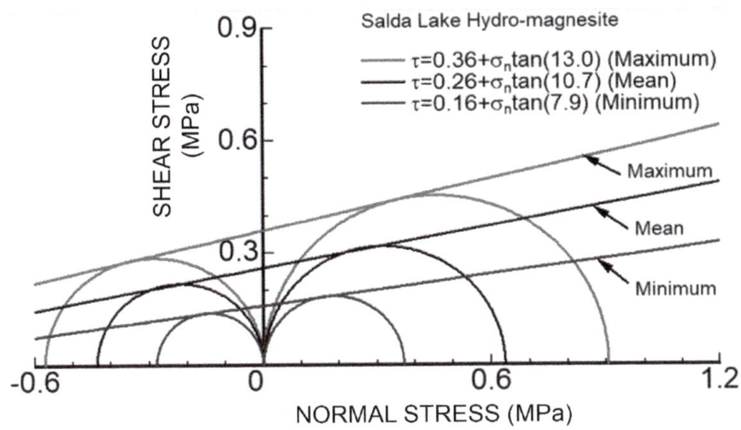

FIGURE 11.10 Mohr–Coulomb strength envelopes.

11.5 IMPLICATIONS ON ROCK AT JEZERO CRATER

The outcomes from the investigations into the characteristics of the rocks of Salda Lake may have important implications for rocks explored by the Perseverance Rover in Jezero Crater on Mars (Figure 11.11). Besides the exploration of some evidences of traces of early life preserved in rocks, the abrasion (grinding), drilling and coring of rocks in Jezero Crater will certainly yield their geo-mechanical characteristics. As the water had already been evaporated in Jezero Crater as a result of the erosion of the atmosphere of Mars due to the weakening of its magnetic field, the rocks in Jezero

FIGURE 11.11 Comparison of Salda Lake and Jezero Crater (modified from NASA 2019).

Crater would be in an almost dry state. Due to large temperature differences as well as strong winds, they may be subjected to some weathering. If the deposits in Jezero crater consist of hydro-magnesite, it is expected that hydro-magnesitic rocks would be quite a bit harder. Some penetration tests with 3 mm flat tip needles were performed on Salda Lake or hydro-magnesite samples to check this hypothesis, and the results are shown in Figure 11.12. As expected, the responses of penetration versus tip pressure differ as the rock hardens with time.

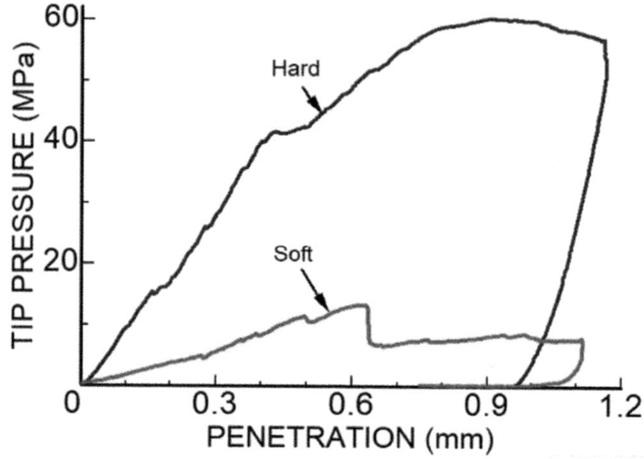

FIGURE 11.12 Penetration versus tip pressure responses of 3 mm needle for hard and soft rocks.

Besides the similarity of the physico-mechanical properties of rocks in Jezero Crater and Salda Lake, it is very likely that some geomorphological features such as toe erosion, cliff failures, rock falls and debris flow along the shoreline, and the adjacent slopes should be observed in Jezero Crater. Furthermore, it is very likely that several shoreline erosion levels may be observed in order to confirm the watery past of Mars.

Comets and Asteroids 12

12.1 COMETS

Comets are considered to be composed of dust, rock, and ice. They range from a few kilometers to tens of kilometers wide. When they orbit closer to the Sun, they heat up and spew gases and dust into a glowing head that can be larger than a planet.

One of the well-known comets is Halley, which is often called the most famous comet. Astronomers have now linked the comet's appearances to observations dating back to more than 2200 years ago, at least. This comet has a return period of 76 years, and it was last seen in 1986. Halley's dimensions are about 15 by 8 km.

Another interesting comet is Comet 67P/Churyumov–Gerasimenko, which was discovered on October 22, 1969, at the Alma-Ata Observatory in Kazakhstan (Figure 12.1). This is the first comet to be orbited and landed upon by the Rosetta spacecraft, carrying the Philae lander of the European Space Agency, from Earth. The mission ended with the spacecraft's controlled impact on the comet's surface on September 30, 2016. Its radius is about 2 km with a density of 533 kg/m^3.

Very interesting observations from the planetary rock mechanics viewpoint were done while orbiting and descending in a controlled impact by the spacecraft on the comet (as shown in Figure 12.2. Figure 12.2 also shows slope failures, rock falls and a sinkhole (skylight) in the comet (Baslilevsky et al. 2017). The observations and their back analyses through analytical methods have provided constraints on the mechanical properties of the cometary material such as tensile, shear and compressive strengths and repose angle, frictional properties (e.g., Groussin et al. 2018; Attree et al. 2018). The observations involve rocky slope, slope debris, boulders, overhanging cliffs and sinkholes. Figure 12.3 shows the estimate of tensile strength of overhanging cliffs using the bending theory of cantilevers (Aydan and Kawamoto 1992; Tokashiki and Aydan 2010; Aydan 1989, 2020a).

12.2 ASTEROIDS

Asteroids are rocky, airless remnants left over from the early formation of our Solar System about 4.6 billion years ago (Figure 12.4). They are also called minor planets, and they orbit within the inner Solar System. The mining of metallic asteroids is now considered by various space agencies. The Hayabusa mission initiated by JAXA Aerospace Exploration Agency was first to bring some materials from the Itokawa asteroid. The second mission, named Hayabusa-2, was successful in bringing back rock-like materials from Ryugu Asteroid (Figure 12.5).

DOI: 10.1201/9781003643852-12

FIGURE 12.1 Image of Comet 67P/Churyumov–Gerasimenko (from ESA 2019).

(a) Toppling type slope failure

(b) Sinkhole (skylight) and rock falls

FIGURE 12.2 Several images of Comet 67P/Churyumov–Gerasimenko (from ESA 2019).

(c) Rockfall

(d) Slope failure

FIGURE 12.2 (Continued)

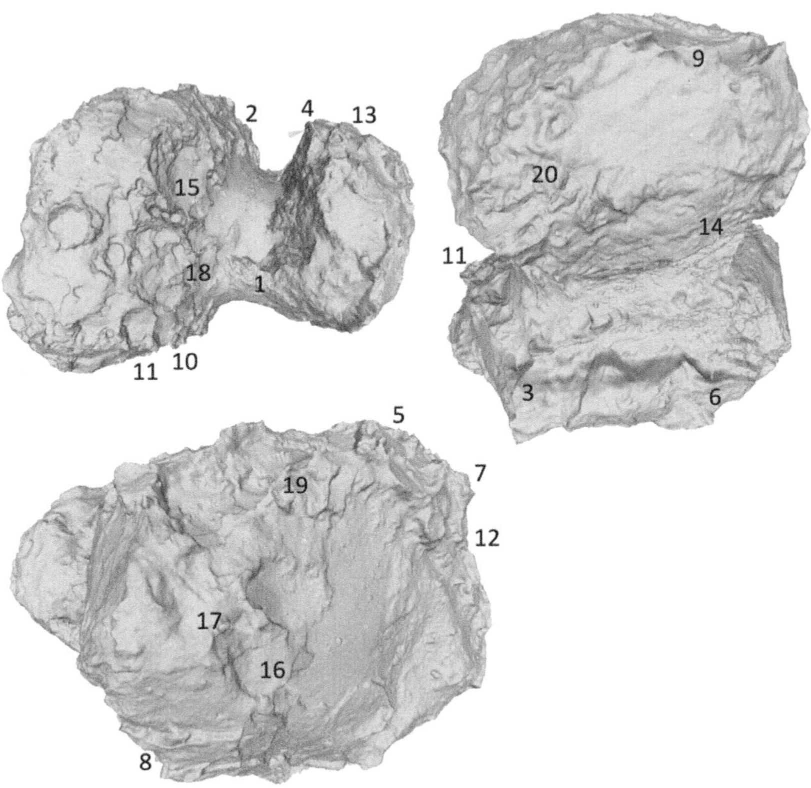

FIGURE 12.3 Estimation of tensile strength of cometary materials from overhanging cliffs of Comet 67P/Churyumov–Gerasimenko (from Attree et al. 2018).

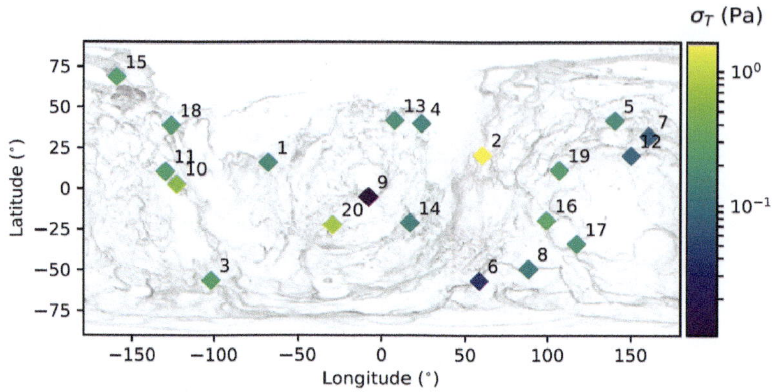

FIGURE 12.3 (Continued)

FIGURE 12.4 View of Pysche Asteroid (NASA).

FIGURE 12.5 Material brought back by Hayabusa-2 (JAXA).

Further Issues in Terrestrial Planetary Rock Mechanics 13

Although humankind has not been on Mars yet, the Mars exploration rovers (MERs: Opportunity, Spirit, Curiosity and Perseverance), Mars Reconnaissance Orbiter (MRO) and InSight Lander have provided tremendous data on various aspects of Mars. Mars Reconnaissance Orbiter (MRO) has provided images on recent events such as impacts by meteorites, mass movements and rockfalls. The InSight Lander has provided data on vibrations induced by seismic events, or marsquakes, and/or by meteorite impacts in addition to other sources. MERs have provided some direct data through drilling, grinding, chemical analyses and imaging on outcrops of rocks on the surface of Mars. Unfortunately, two of the Rovers, Opportunity and Sprit, lost their power units due to prolonged dust storms on Mars, and they are no longer functioning. On the other hand, the power of Curiosity is based on nuclear energy, and it has been currently roving more than 36 km (as of July 2024) and has provided tremendous valuable data on the rocks of Mars in Gale Crater. Perseverance has been also providing similar data in addition to data from onboard seismic sensors (NASA 2022).

13.1 VIBRATIONS DUE TO METALLIC WHEELS OF ROVERS AND THEIR EFFECTS

When one closely examines the images from the MERs, one can easily notice the effects of vibrations induced by their metallic wheels as seen in some of the images released by NASA (2022). The recent video released about the sounds caused by the Perseverance clearly indicated some shock-like vibrations when the rovers drove over rock fragments. The effects of the vibrations caused by the metallic wheels are quite high on hilly ground and overhanging rock blocks, layers and some surficial sliding and block sliding/falls occurred. Figure 13.1 shows examples. It seems that the vibrations caused by metal wheels of MERs during roving are probably amplified. Figures 13.2 and 13.3 show the amplification of motions on the top of model slopes with/without toe erosion subjected to base acceleration during sweeping tests in shaking table tests (Aydan 2021a, 2022b, 2023b). As noted from these figures, the vibrations may be amplified up to 4–5 times that at the base.

DOI: 10.1201/9781003643852-13

FIGURE 13.1 Sliding or fall of rock blocks and surficial sliding.

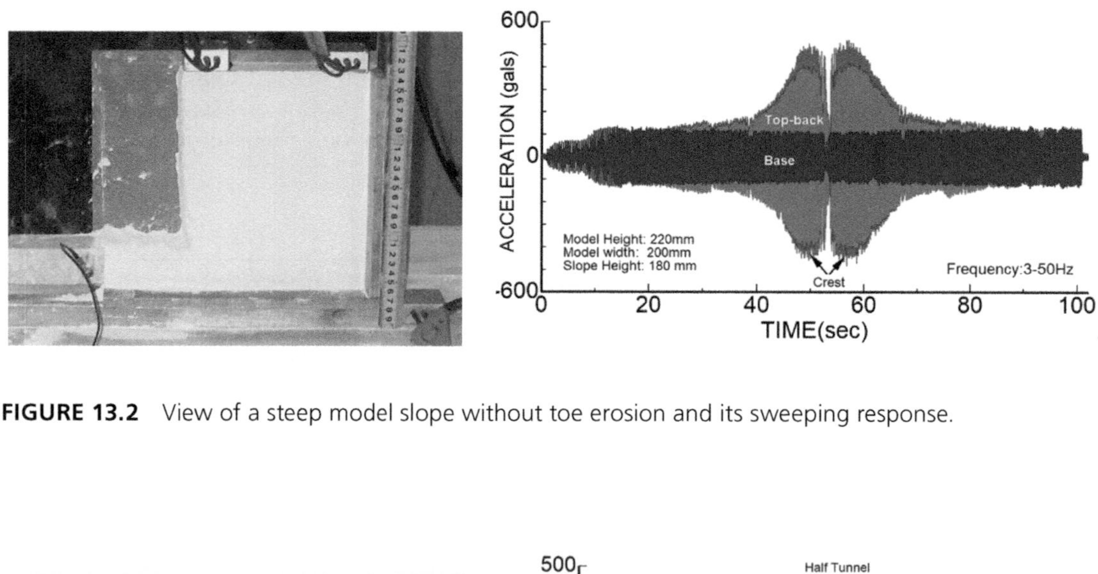

FIGURE 13.2 View of a steep model slope without toe erosion and its sweeping response.

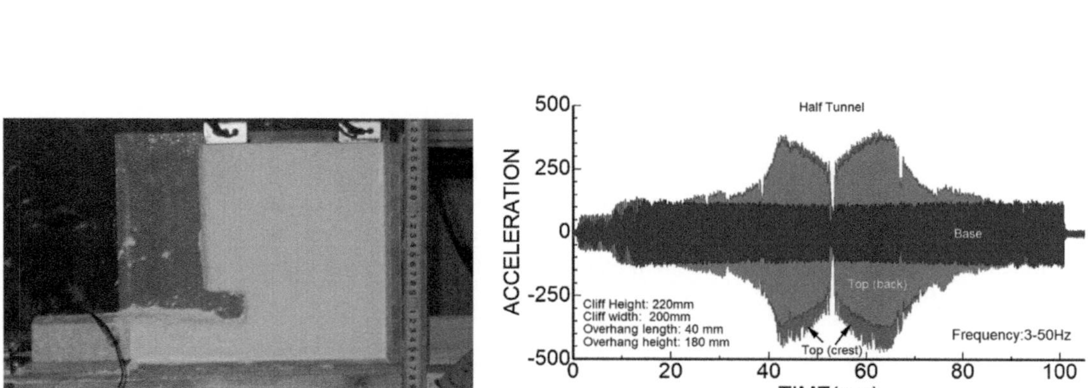

FIGURE 13.3 View of a steep model slope with toe erosion and its sweeping response.

13.2 DRILLING- AND GRINDING-INDUCED VIBRATIONS AND THEIR EFFECTS

The MERs are equipped with drilling and grinding tools for investigating rock surfaces and rock blocks (Aydan 2023b). During operations, the drilling and grinding processes induce some vibrations, and they cause either fracturing and/or displacing rock blocks. The soil deposits on the target rock block surfaces are also displaced or sink along existing rock fractures. Sometimes these observations are wrongly interpreted as soil liquefaction in Mars. It was simply a surficial sliding of granular material and the appearance of non-weathered material beneath. Figure 13.4 shows such effects of the vibrations-induced during drilling/grinding operations. In addition, the weight of the MERs caused some fracturing of thin platy or breakage of elongated rock pieces due to bending stresses.

FIGURE 13.4 Effects of drilling/grinding operations on target rocks and their surroundings.

Vesicular-type basalt is commonly observed on various parts of Mars. Although basalt is known to be very high-strength rock, its porous structure causes the overall reduction of the strength, as shown in Figure 10.4. The samples from Mt. Fuji, Japan, Aswan in Egypt, Aksaray in Türkiye and Wanda Mines near Iguassu Falls in Argentina were utilized for measuring the vibrations caused during drilling and grinding (Figure 13.5). As examples, vibration measurements on vesicular basalt samples from Mt. Fuji were selected, and some vibrations measurements grinding/drilling operations were recorded. Figures 13.6 and 13.7 show some acceleration records obtained from the stand-alone type QV3-OAM-SYN accelerometer during grinding/drilling operations. The surface was vertical and accelerations were measured parallel and perpendicular to the surface of the grinding/drilling and vertical direction of the block. Regarding the grinding operation, the largest acceleration occurred perpendicularly to the grinding surface, and it can be up to 2200 gals, while the other components were about 1500 gals. As for drilling operations, the maximum accelerations were less than 900 gals, and the largest acceleration occurred perpendicularly to the drilling surface.

13.3 PROPERTIES OF ROCKS FROM OTHER TERRESTRIAL PLANETARY BODIES

As explained in the previous sections, numerous experiments on rock, rock discontinuities and rock masses have been carried out in the field of Rock Mechanics and Rock Engineering, and many mechanical and engineering properties have been determined. Nevertheless, it is needless to say that similar

FIGURE 13.5 Views of samples and mounted bits.

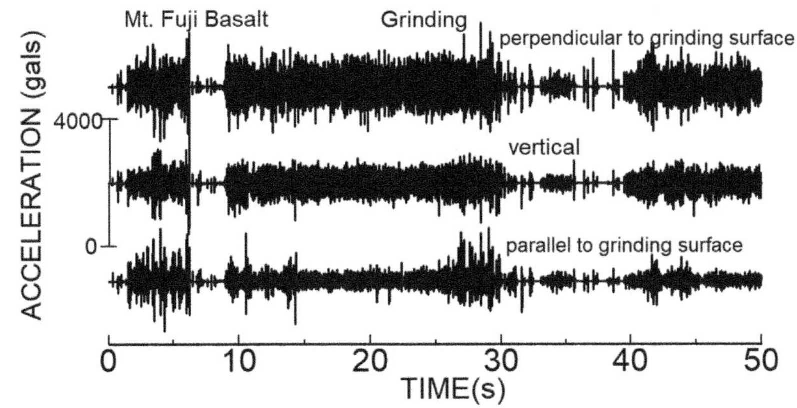

FIGURE 13.6 Records of accelerations induced during grinding operation.

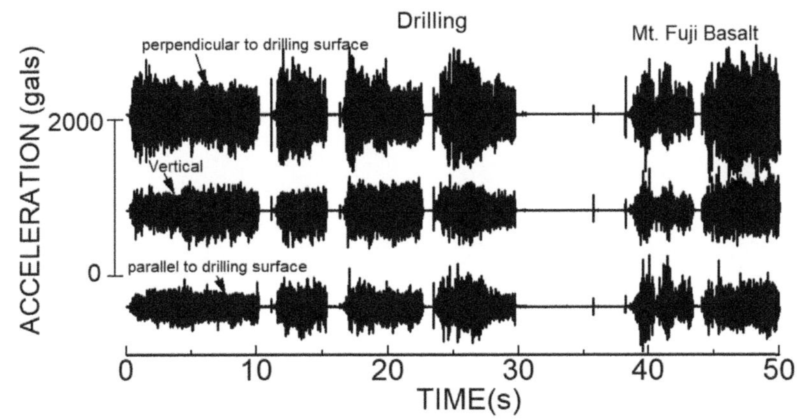

FIGURE 13.7 Records of accelerations induced during drilling operation.

experiments for extraterrestrial rocks should be carried out probably in their own environment. During the Apollo program, more than 300 kg of rock sample were brought to Earth. However, no mechanical experiments have been carried out yet. The Mars rovers provided many images of rocks, rock masses and discontinuities. Back analyses of many images (some examples are shown in Figures 13.8 and 13.9) may provide information on the mechanical properties of rocks.

13.3.1 Inference of Properties from Rocks Broken by MERs

As shown in Figure 13.8, a platy rock piece was broken due to bending stress under the weight of the MER through its metallic wheel. This type of breakage can be utilized to infer the bending (tensile) strength of terrestrial planetary bodies. The bending theory for beams on elastic foundations may be utilized for such a purpose. If the loading is somewhat complicated, numerical analyses may be used. The simple technique might be the utilization frame type analyses capable of considering ground reaction.

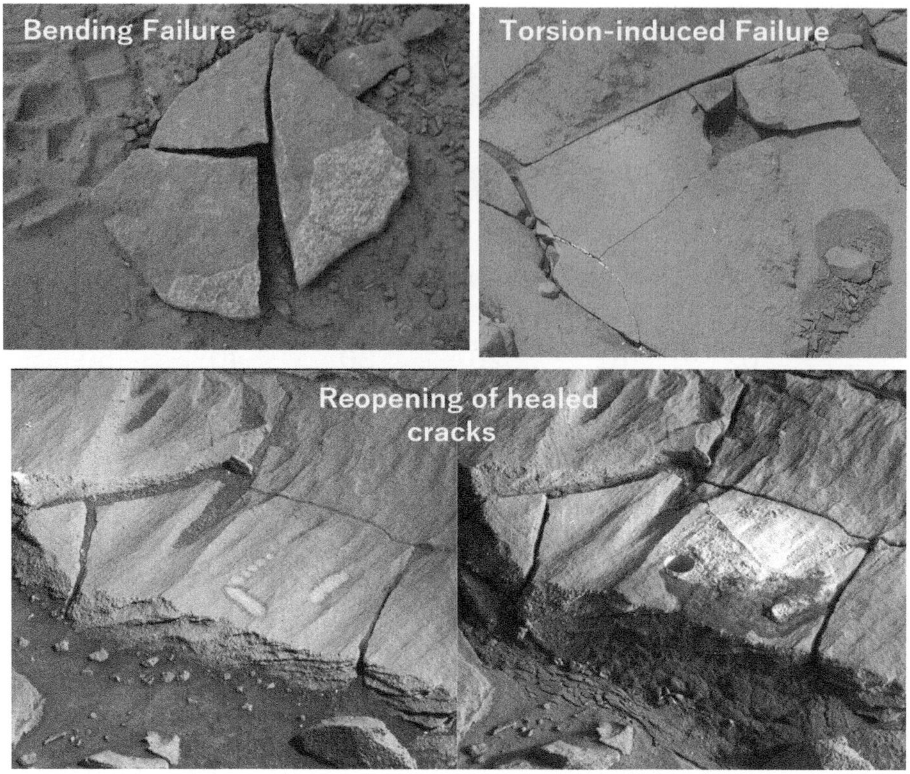

FIGURE 13.8 Broken or fracture platy rock fragments due to MERs.

A finite element method capable of considering elastic, elasto-plastic constitutive laws can be used. Such an example is shown in Figure 13.9. In this particular analysis, a 10 cm thick slab is subjected to a load level of 30 kN. As seen from Figure 13.9, tensile stresses develop at the lower side of the slab, which explains why the rock plate shown in Figure 13.8 was broken by the metallic wheel resulting from the weight of the MER. This example would provide, at least, the tensile strength of the rock on Mars.

13.3.2 Inference of Properties from Drilling and Grinding Operations by MERs

The force and rate of advance during drilling operations by the MERs, such as Curiosity or Perseverance, may also provide another data set on the mechanical properties of intact rocks (Aydan 2022b, 2023b). Particularly, the utilization of drilling performance and drill hole tailings shown in Figure 13.10 and drilling performance may yield additional data on the physico-mechanical properties of rocks. Besides empirical techniques, numerical methods may be utilized to evaluate drilling performance and back analyses of the physico-mechanical properties. Figure 13.11 shows a simulation of a drilling-type excavation using the finite element method. In this simulation, the stress states in the vicinity of the drill hole are evaluated for excavation depths of 0, 5 and 10 cm, and the tip pressure and torsion acting on drill bits are taken into account.

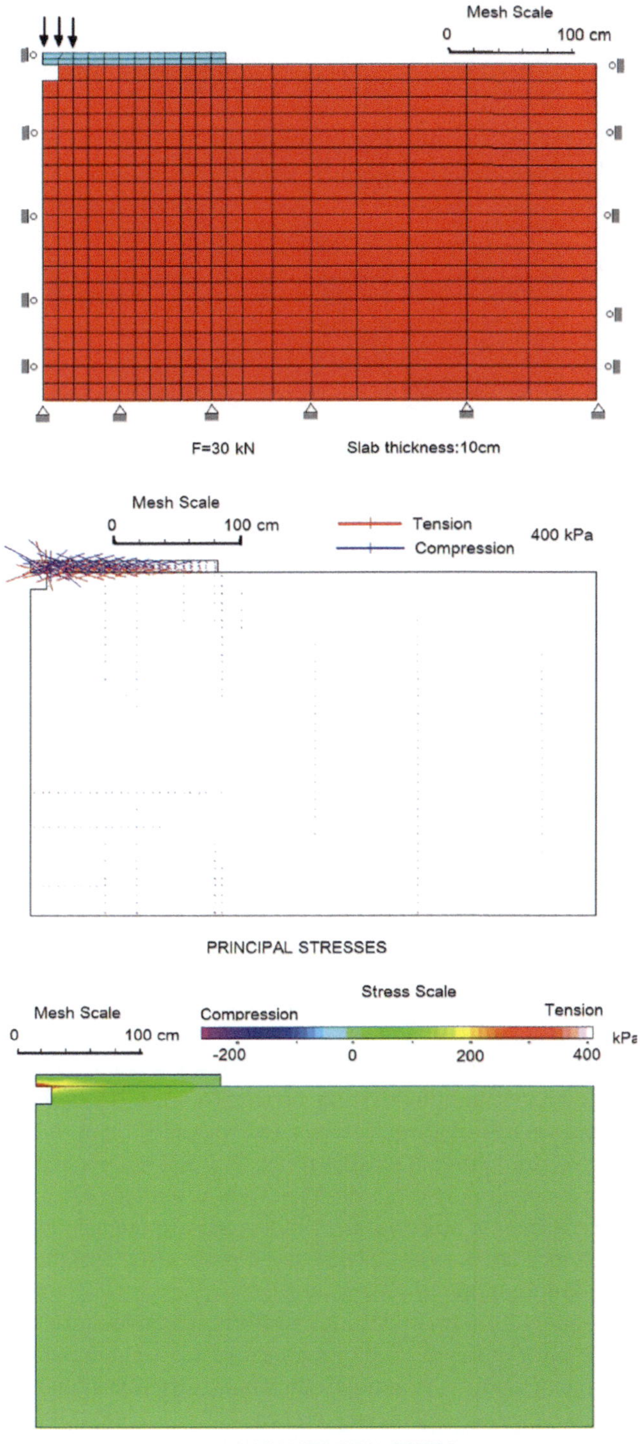

FIGURE 13.9 Mesh and computed principal stresses and principal stress contour in the vicinity of a slab subjected to a load level of 30 kN by the finite element method.

(a) Mars (b) Earth

FIGURE 13.10 Views of drilling tailings on Mars and in the laboratory on Earth.

13.3.3 Inference of Properties from Failed Slopes

One may easily infer the frictional properties from back analyses of stable and unstable blocks from the base inclination and repose angles, although it may be slightly lower from the intrinsic friction angle. Furthermore, the back analyses of failed rock slopes and rock cliffs may further provide on strength properties of rock masses such as cohesion, tensile strength for a given friction angle (Figures 13.12 and 13.13). Such analyses have been already reported by several researchers (e.g. Lucchitta 1979; Schultz 2002; Neuffer and Schultz 2006) for slope failures in terrestrial planetary bodies. It should be noted that the failure modes of slopes are various, as discussed previously, and that the analysis techniques differ depending upon the failure mode.

Figure 13.14 shows a model for dynamic limit equilibrium analyses of a rock slope against planar sliding (e.g., Aydan 2015), and Figure 13.15 shows an example of a computation for an actual earthquake record. The latter type of analysis technique can handle different situations relevant to the conditions of such examples in the respective planetary bodies.

Figure 13.16 shows some back analyses of wedge sliding failure in model testing by considering the variation of friction and wedge angle under gravitational and seismic loading (e.g., Kumsar et al. 2000; Aydan and Kumsar 2010). Figure 13.17 shows some examples of back analyses of the dynamic response of model wedges subjected to seismic loads.

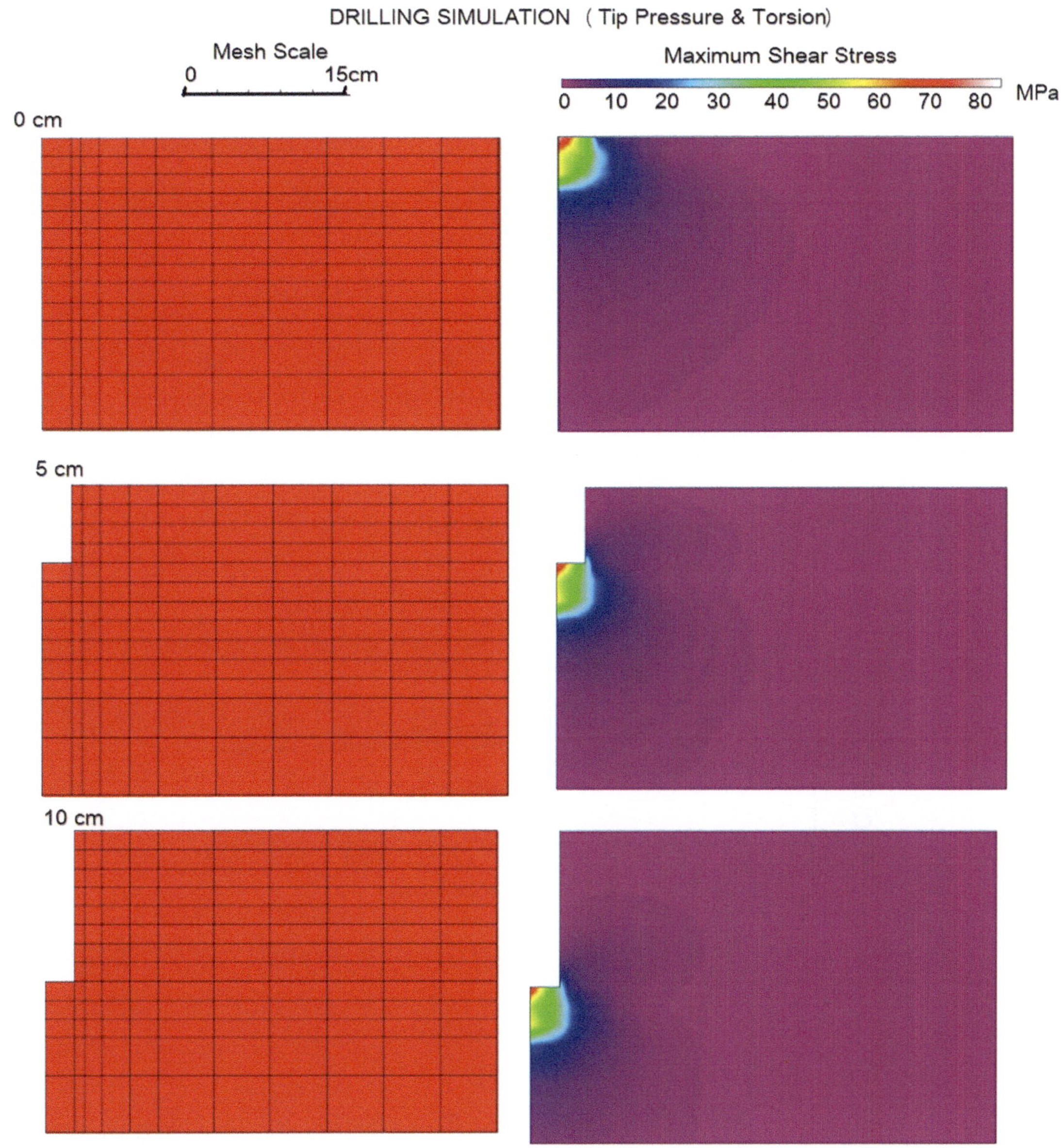

FIGURE 13.11 Mesh configurations and maxim shear stress distribution in the close vicinity of drill hole for excavation depths of 0, 5 and 10 cm.

Besides static and dynamic limiting equilibrium methods, numerical methods can also be utilized. However, it should be noted that the consideration of discontinuities in rock mass in numerical analyses would be necessary. There are different techniques for this purpose. For example, the discrete finite element (DFEM) technique can be used for such a purpose. This method considers discontinuities in rock mass and large deformations using the updated Lagrangian scheme. Several examples of the utilization of this method for slopes subjected to planar and circular sliding and flexural toppling failures are shown in Figures 13.18, 13.19 and 13.20 (e.g., Aydan et al. 1996a, 1996f; Aydan 2017b, 2021c). The details of analyses can be found in the respective references.

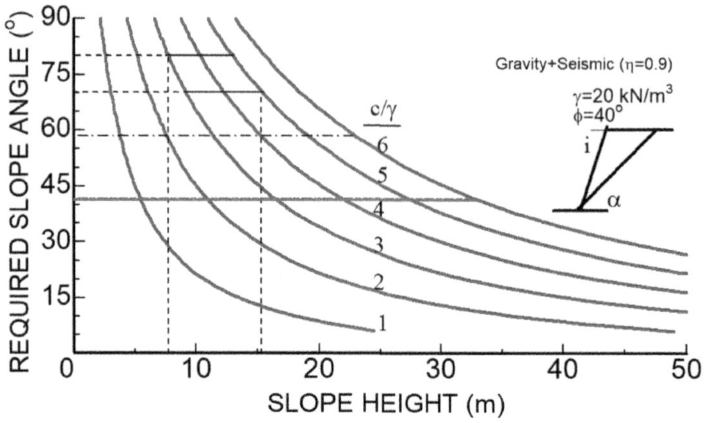

FIGURE 13.12 Inference of cohesion for a planar sliding failure of rock slope.

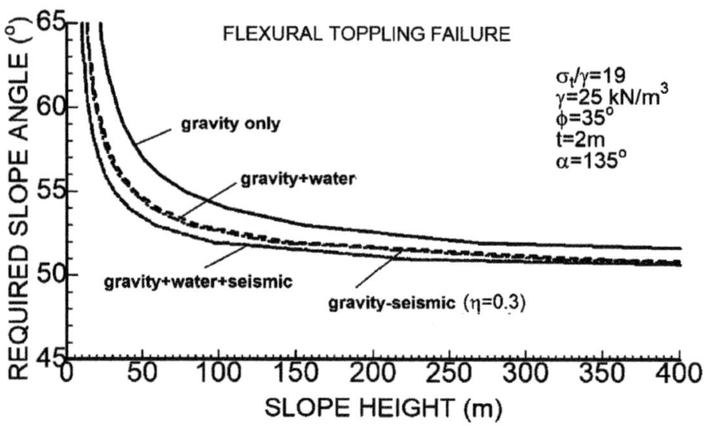

FIGURE 13.13 Assessment of a rock slope against flexural toppling failure under various conditions using the method of Aydan and Kawamoto (1992).

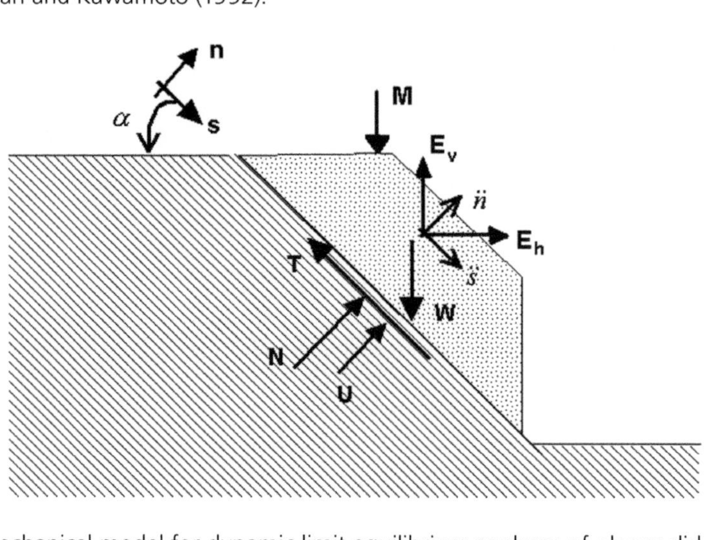

FIGURE 13.14 Mechanical model for dynamic limit equilibrium analyses of planar slides.

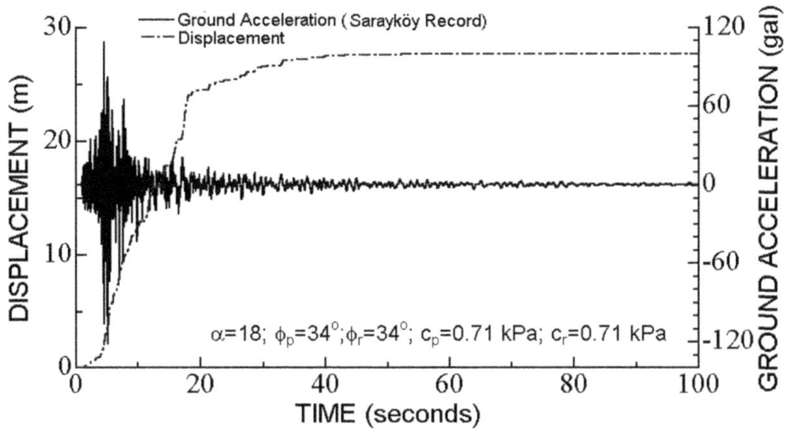

FIGURE 13.15 Computational example for the response of a slope during planar sliding.

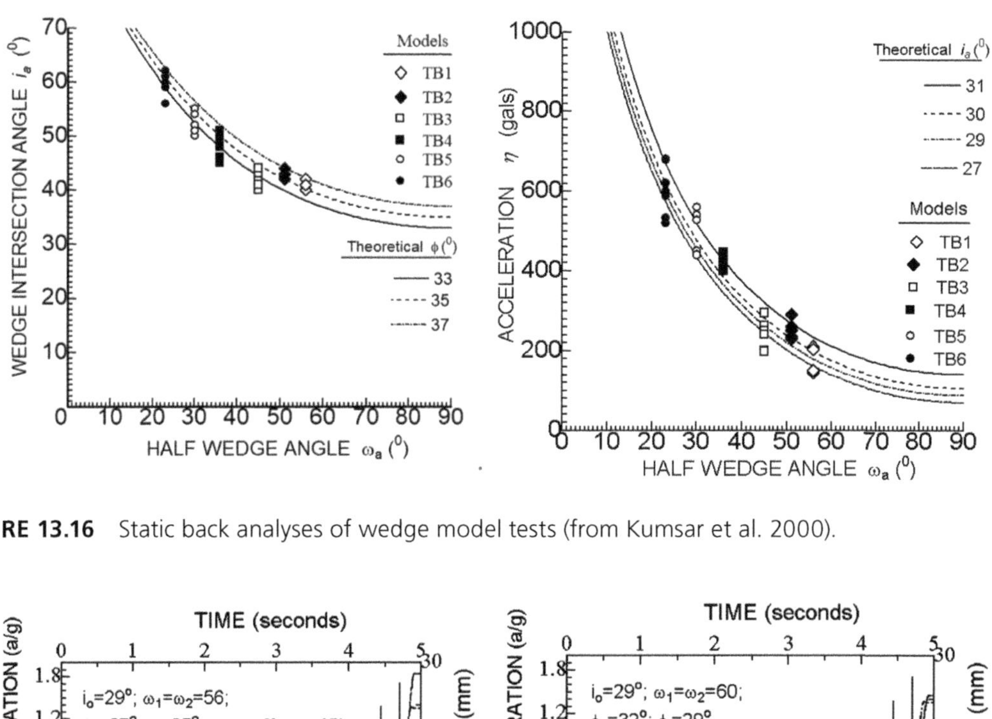

FIGURE 13.16 Static back analyses of wedge model tests (from Kumsar et al. 2000).

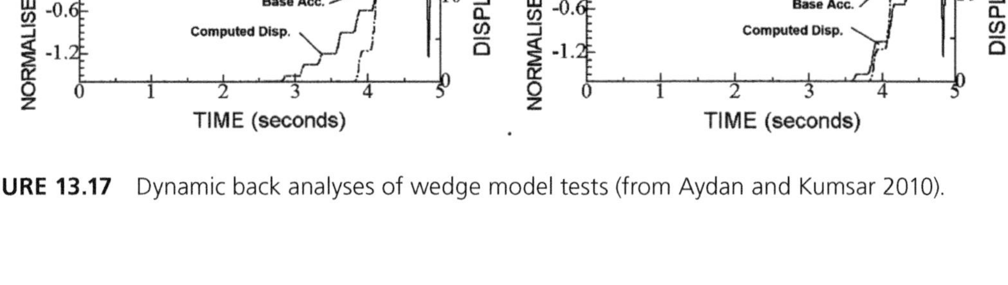

FIGURE 13.17 Dynamic back analyses of wedge model tests (from Aydan and Kumsar 2010).

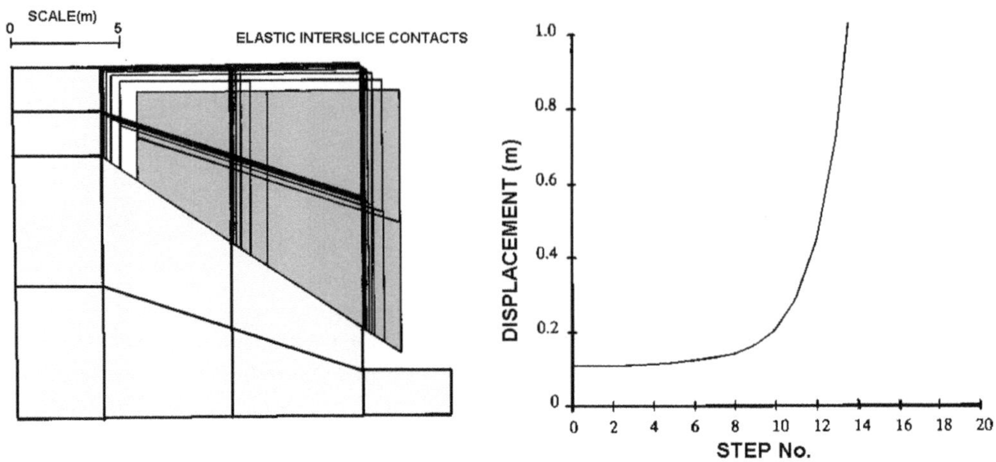

FIGURE 13.18 DFEM analysis of a slope subjected to planar sliding.

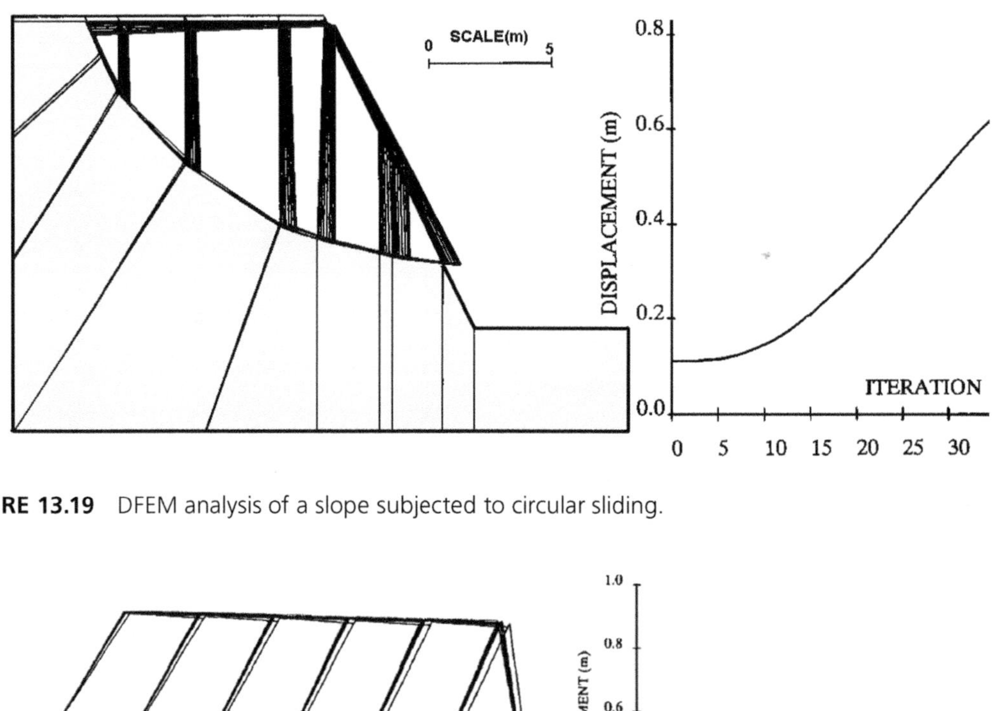

FIGURE 13.19 DFEM analysis of a slope subjected to circular sliding.

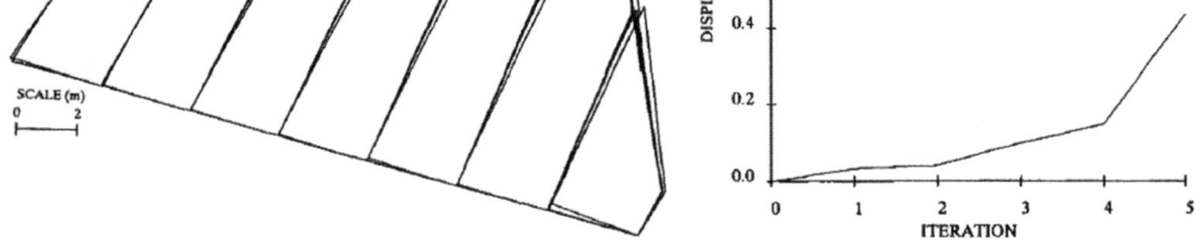

FIGURE 13.20 DFEM analysis of a slope subjected to flexural toppling.

13.4 INFERENCE OF FRICTIONAL PROPERTIES OF DISCONTINUITIES

Tilt tests are often used to infer the frictional properties of saw-cut surfaces as well as natural rock discontinuities (e.g., Alejano et al. 2018; Aydan et al. 2023). Figures 13.21 and 13.22 show the mechanical model and views of some tilt test device. Tilt angle measured during the tests will correspond to the static and kinetic friction angle of discontinuities if appropriate instrumentation is employed (e.g., Aydan et al. 2023). The images taken during the explorations by MERS in the Mars and Apollo landings on the Moon by NASA and the Venus landing by the Venera Program of the former Soviet Union should yield some frictional properties of discontinuities and regolith in respective planetary bodies with appropriate mechanical considerations. Figure 13.23 shows several pictures of rock mass conditions and rock blocks on inclined surfaces and discontinuity planes in Mars. For example, one can infer the friction angle of discontinuities and/or block–mass interface ranging between 35 and 42 degrees from examples shown in Figure 13.23, which are also very close to those on Earth. However, the displaced blocks may also yield some kinetic friction angles or ground motion data in the case of seismic events. Such an example has been already discussed in subsection 9.1.9.

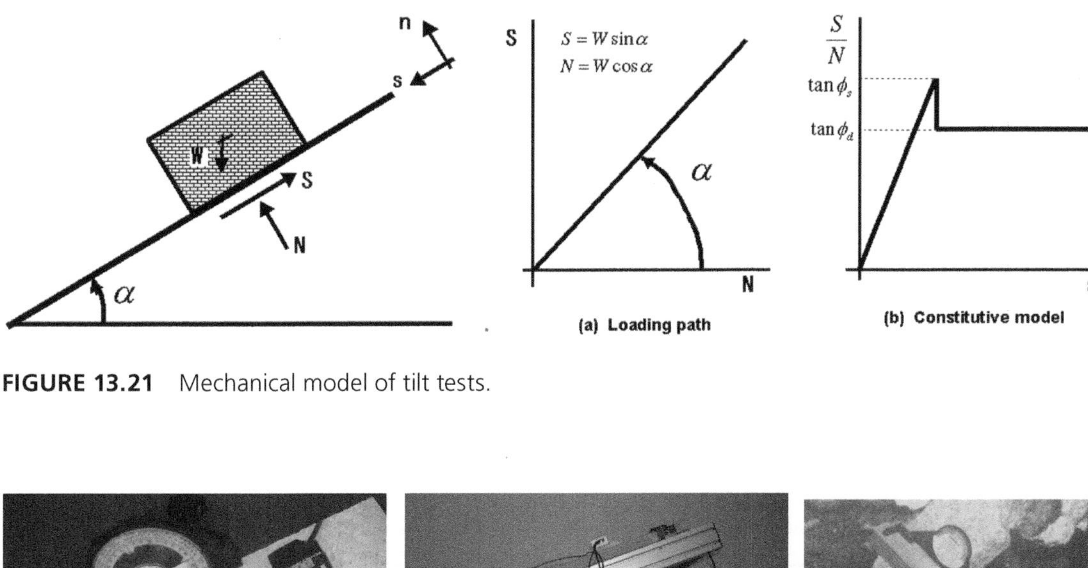

FIGURE 13.21 Mechanical model of tilt tests.

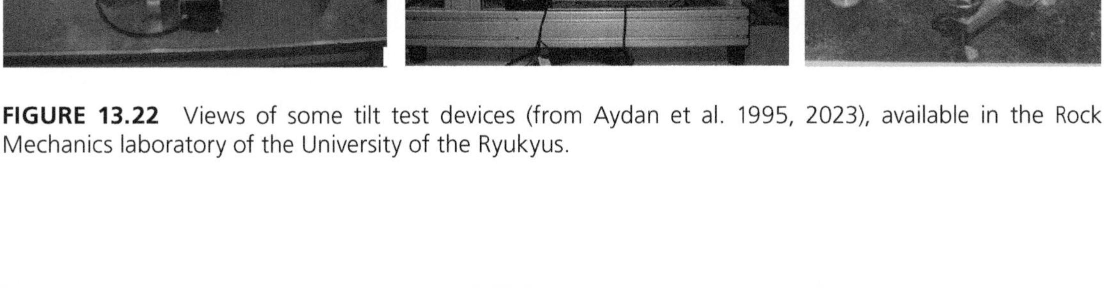

FIGURE 13.22 Views of some tilt test devices (from Aydan et al. 1995, 2023), available in the Rock Mechanics laboratory of the University of the Ryukyus.

Magnarini et al. (2023) described a large mass movement near the landing site of Apollo 17. The estimated reach angle for this site was about 12.9 degrees where the mountain slope had an inclination of 21.2 degrees. The reach angle for this site is very close to what happened at the Ontake mountain failure caused by the 1984 Nagano-Seibu earthquake with a magnitude of 6.8.

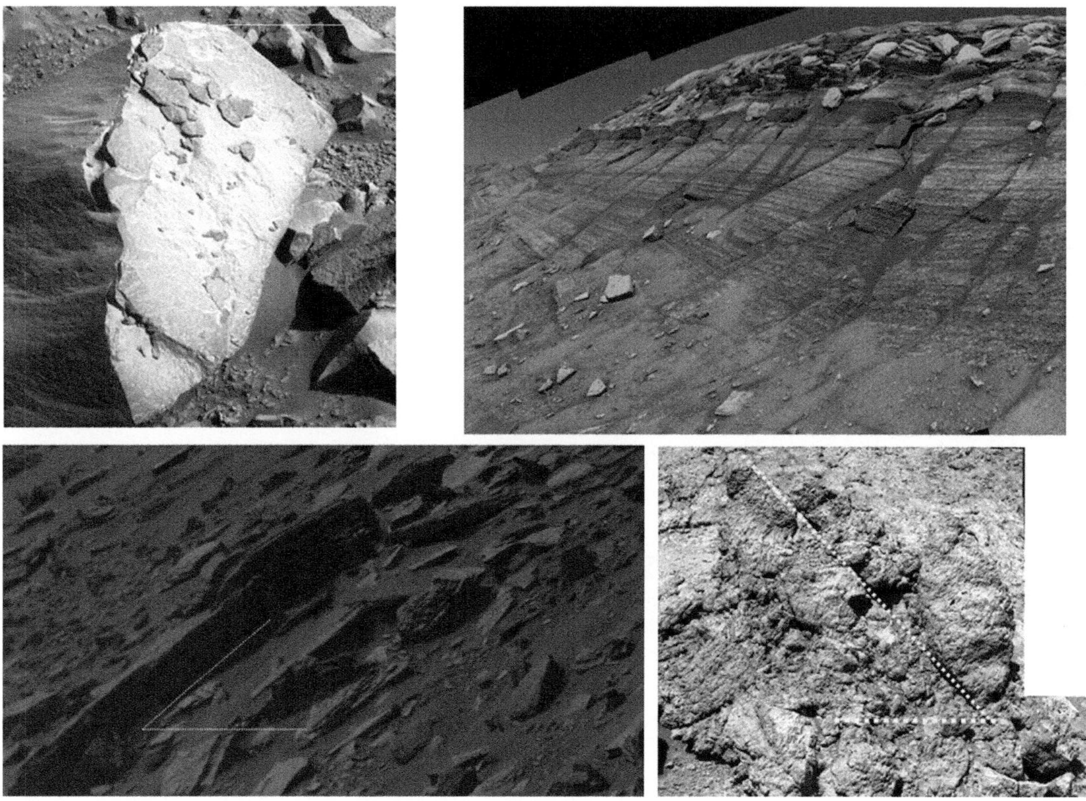

FIGURE 13.23 Images to infer frictional properties of discontinuities on Mars.

13.5 UTILIZATION OF INDEX TEST TECHNIQUES

Many index test techniques have been developed in the field of Rock Mechanics and Rock Engineering. Although the utilization of proper testing techniques is always necessary, it is almost impossible to implement them during explorations of rocks and discontinuities in other terrestrial planetary bodies. Therefore, this section addresses this issue and describes some index methods for assessing the physico-mechanical properties of rocks, available in the field of Rock Mechanics and Rock Engineering (e.g., Aydan 2012b; Aydan et al. 2014a; Ulusay et al. 2014, 2025; Aydan and Ulusay 2024), which may be utilized in the next-generation exploration rovers. In addition, numerical studies require the evaluation of the parameters of constitutive laws, and some available techniques for such a purpose have been introduced in this subsection.

13.5.1 Elastic Wave Velocity

When rocks are assumed to be isotropic, P-wave and S-wave velocities of samples are measured in one direction. However, they should be measured perpendicularly to bedding (schistosity) and two mutual directions parallel to the bedding plane (schistosity) if the anisotropy is dominant. Such examples are described in the literature (e.g., Aydan and Ulusay 2003, 2013; Aydan et al. 1993, 1996b; Sezaki et al. 1990, 1994; Seiki and Aydan 2003), and there are many empirical relations to establish the relations among wave velocities and strength and deformability characteristics. These type of measurements may also be done by equipping the explorers with appropriate devices. If humankind gets access to the exploration sites, similar devices can also be used in such explorations of the terrestrial planetary bodies.

13.5.2 Schmidt Rebound Number

The Schmidt hammer, which was initially developed for the concrete of civil and architectural structures, is used in Rock Mechanics and Rock Engineering applications with some modification of the devices (Seiki and Aydan 2003; Aydan and Ulusay 2024). Schmidt hammer can also be used as an index testing device in planetary explorations. Therefore, the present exploration rovers may be equipped with such devices. However, the major issue would be the interrelations between the rebound value and strength and deformability properties as the gravitation fields in planetary bodies are different. When this index method is utilized, some consideration on the existence of inclusions in rock matrix should be also given as the rebound values will be different for soft/hard matrix and hard/soft inclusions. In addition, the direction of measurement must be also taken into account.

13.5.3 Needle Penetration Index Tests

The needle penetration device is developed by the Maruto Co. of Japan (Maruto Co. 1999). The device can measure the applied load up to 100 N, and the penetration depth is up to 10 mm. The diameter of the needle is 0.82 mm (Ulusay et al. 2014; Aydan 2012b; Aydan et al. 2014a). The ratio of applied load to penetration depth is called the needle penetration index (NPI). The original motivation to develop this device was to assess the properties of squeezing rocks (soft rocks with UCS up to 30 MPa) in tunneling and soft rocks beneath large bridge foundations, as it is often difficult to sample rocks and to test them in laboratory. The damage caused to the rock by the needle penetration is also quite negligible, and the device can also be effectively used in assessing the state and properties of rocks in archeological remains excavated in/on rock mass, particularly.

Another penetration test device utilizing a low-capacity loading device and a special rod-like loading platen with a flat end was developed by Aydan et al. (2008). The device is presently used in laboratory conditions. However, it can be easily used in field explorations. The diameter of rod-like platen ranges between 1 and 3 mm. However, experimental results indicate that there are some undesired stress concentrations when the diameter is less than 1 mm, and the rod-like platen may buckle when high-strength rock is tested. For the interpretation of the experimental results, the nominal strain is defined in terms of penetration displacement divided by rod diameter. Furthermore, the applied stress is defined by dividing the applied load by the area of the flat end. The device is equipped with a load cell and a displacement transducer to record load-displacement responses. In addition, an AE sensor is attached to the sample to monitor the cracking phenomenon. This device has also big potential to be utilized in explorations in terrestrial planetary bodies.

13.5.4 Equotip or Leeb Hardness (Rebound) Tests

Equotip hardness (rebound) test device was originally developed for evaluating the hardness of metals. However, since the 1990s, it is also widely on rocks (e.g., Aydan and Ulusay 2024; Ulusay et al. 2025). The device measures the ratio of a rebound height of a small hardened steel sphere to the original drop height, and the ratio is multiplied by 1000. As the test area is very small, it can be utilized in many places, and it is very likely to measure also the state of matrix and inclusions in some rocks. It is probably one of suitable techniques for terrestrial explorations as it is a small device, and data storage should be quite easy with some small technical modifications. It should be noted that some slight damage always and definitely occurs beneath the impact locations, even in many hard rock samples, and the Equotip hardness (rebound number) differs even on the same surface and the surface conditions.

13.6 INFERENCE OF PARAMETERS OF ELASTO-PLASTIC CONSTITUTIVE LAW

The analytical and numerical methods require the parameters of constitutive laws, which are not strength parameter only. In other words, some researchers mistakenly consider the constitutive laws to be only strength criteria. The determination of the fundamental parameters of constitutive laws for rocks constituting terrestrial planetary bodies is almost impossible without actual rock samples. In this subsection, the procedure for determining some important parameters based on uniaxial compression strength (UCS) is introduced. Figure 13.24 shows the idealized strain–stress relations during deformation, together with internal fracturing of rocks. As shown in Figure 13.25, Aydan et al. (1993, 1996b) defined some plastic deformation parameters using the strain levels for post-yielding behavior.

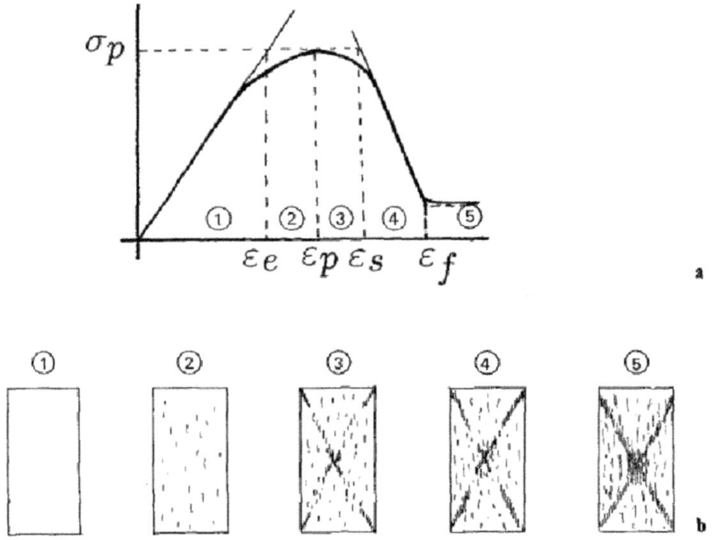

FIGURE 13.24 Definition of plastic deformation parameters using the strain levels for post-yielding behavior.

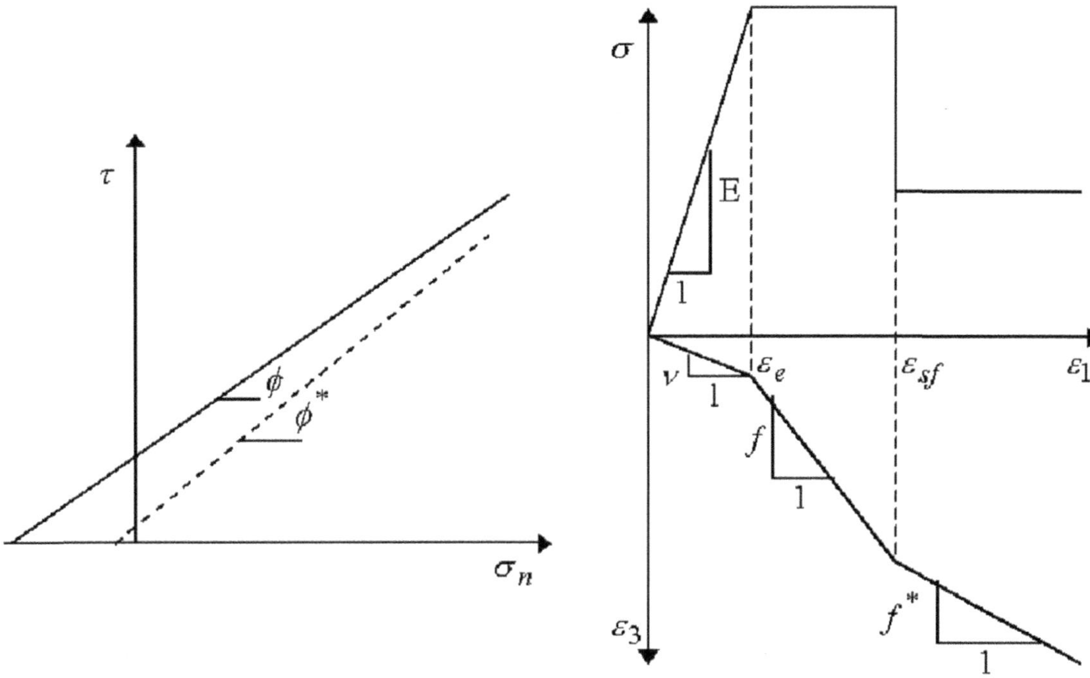

FIGURE 13.25 Idealized strain–stress relations during deformation.

The details of the parameters and their utilization have been described in the publications of Aydan et al. (1993, 1996b), which were initially developed for rocks of tunnels subjected to squeezing phenomenon, and it was extended to high-strength rocks subjected to rock burst conditions. These parameters were utilized to assess the deformation responses of various rock structures on Earth. It is also common not to find many parameters used in analytical and numerical analyses. Even for such cases, simple parameters such as uniaxial compressive strength and/or elastic wave velocity can be utilized. Sometime, elastic-wave velocity is more easily found. Therefore, the interrelation between elastic wave velocity and uniaxial compressive strength, shown in Figure 13.26, may be utilized if there are no data on uniaxial compressive

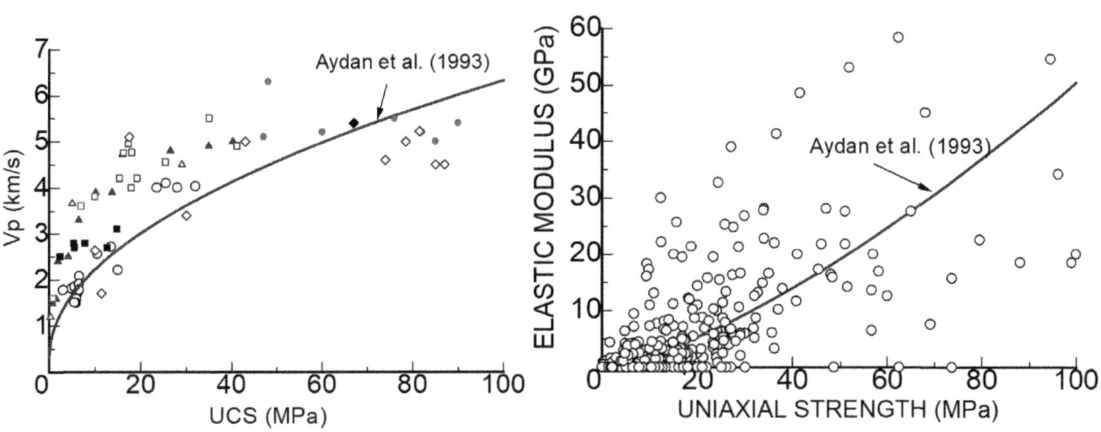

FIGURE 13.26 Interrelations uniaxial compressive strength (UCS) among P-wave velocity and elastic modulus.

strength (UCS). However, if the uniaxial compressive strength (UCS) data are available, some of the direct relations established by Aydan et al. (1993, 1996b) and extended to high-strength rocks (Aydan et al. 2001; Aydan 2018a), shown in Figure 13.27, can be utilized to evaluate the parameters of post-yielding of rocks to assess the response of rock structures in other terrestrial planetary bodies using the analytical and numerical analyses. The same methodology may also be extended to evaluate the visco-elasto-plastic parameters of rock in other planetary bodies using the knowledge gained from the rocks of Earth.

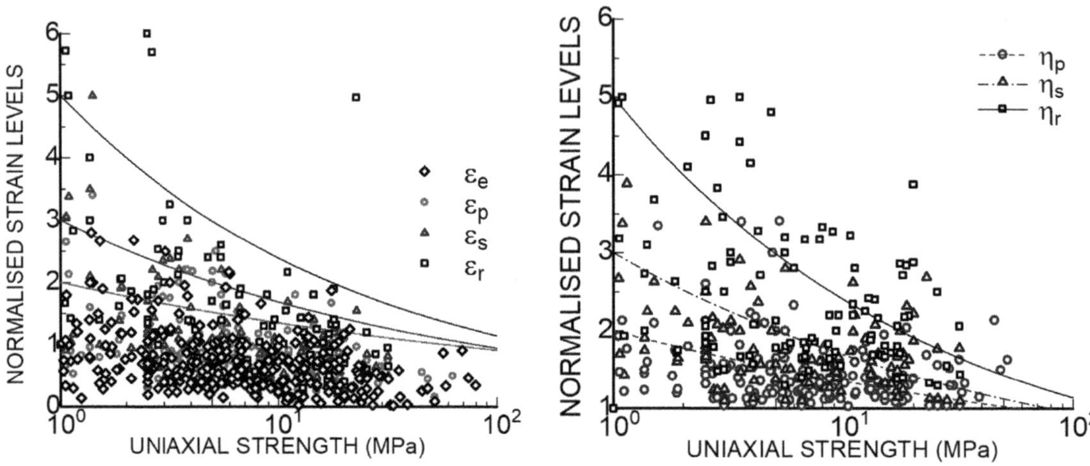

FIGURE 13.27 Interrelations uniaxial compressive strength (UCS) among plastic behavior parameters.

Settlement and Exploitation in Terrestrial Planetary Bodies

14

Mankind exploring the ways to settle on other terrestrial planetary bodies and the activities in US and former Soviet Union in the 1950s may be counted as specific attempts. Both the US and the former Soviet Union sent various unmanned and manned spacecrafts to space. Yuri Gagarin of Soviet Union was the first person to journey into outer space. The humankind landed on the Moon during the Apollo Program, and there have been many attempts to explore Mars and Venus. The US was successful in putting exploration rovers on Mars and exploring it since 1997 (Spirit and Opportunity rovers were the Exploration Rovers on Mars in 2004). The main goals of the space activities may be (e.g., West and Clarke 2010; Aydan 2020b):

1. Exploration of space and terrestrial planets for purely scientific purposes,
2. Mining and exploitation of natural resources;
3. Establishing bases on nearby terrestrial bodies for deeper space exploration.

This chapter is concerned with summarizing the current situation and future trends.

14.1 SETTLEMENTS

the Moon is the closest planetary body near Earth, and it is a target of settlement. The Moon is reachable from Earth in three days, and it has a near-instant communication to Earth. The Soviet Union sent a spacecraft, Luna 1, in 1959 to the Moon in a flyby mission and sent several Luna spacecraft to the Moon; Luna 9 was the first spacecraft to land on the Moon. The US was successful in landing manned spacecraft on the Moon surface. During the Luna missions, the Soviet Union was also able to gather Moon soils. The Soviet Union had a project named Zvezda (a lunar base in Russian), and it at first intended to establish a settlement on the Moon after 1962. This project was abandoned in 1976 following the failure of launches between 1969 and 1972.

The US initiated the Apollo program in 1961 and landed the first humans on the Moon in 1969. Figures 14.1 and 14.2 show the Apollo 11 Lunar Module and the first step of humankind on the Moon. Except for the Apollo 13 mission, Apollo 12, 14, 15, 16 and 17 missions were all successful. The first seismic instrument was installed on the Moon surface. The Apollo 14 Lunar Landing Module landed

DOI: 10.1201/9781003643852-14

FIGURE 14.1 Apollo 11 Lunar Landing Module (Eagle) (from NASA).

in the lunar highlands. The rock samples from the Apollo 14 landing site, named Fra Mauro, were classified as breccias. Astronauts of Apollo 15 collected 77 kg rock and soil samples. Seismometers were installed during the Apollo 14 and 15 missions. During the Apollo 16 mission, the fourth seismic station was installed. The Apollo 17 mission landed at Taurus-Littrow with a large slope failure and debris flow, and the rock samples amounted to 115 kg.

In addition to the landings on the Moon by the US, several unmanned landings have been recently achieved by Japan, India and China. Probably the successful landings on the Far side of the Moon by China could bring soil and rock samples from the Moon to Earth.

The initial "settlements" of humankind on the Moon have been temporary so far. Nevertheless, the USA initiated a new manned lunar program in 2017, the Artemis program. The program's stated long-term goal is to establish a permanent base on the Moon to facilitate human missions to Mars. It is intended to reestablish a human presence on the Moon for the first time since the Apollo 17 mission in 1972 and to continue the direct exploration of Mars begun with data from the Mariner 9 probe in the same year.

The main goal of the settlements on the Moon is thought to be minable minerals under no atmosphere and low gravity, which can make it extremely easy to ship materials and products to orbit. Abundant ice trapped in permanently shadowed craters near the poles could provide support for the water needs of lunar settlements.

As for Mars, there are no settlements except the landing of the Mars Exploration Rovers, so far.

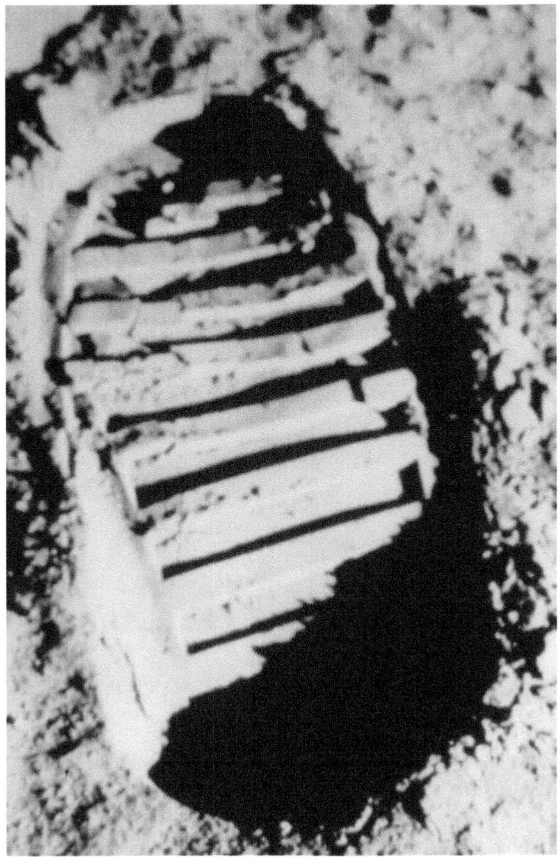

FIGURE 14.2 Imprint of the first step of humankind on the Moon (from NASA).

14.2 EXPLOITATION

Native precious metals, such as gold, silver, and probably platinum, are said to be concentrated at the lunar poles due to electrostatic dust transport. Only a few materials on the Moon have been identified to make economic sense to ship directly back to the Earth, which are helium-3 (for fusion power) and rare-earth minerals (for electronics). If mining is going to employed on the Moon and Mars, the first approach is likely to be open-pit mining. Figures 14.3 shows a possible open-pit mine, which would definitely require high energy consumption. Without nuclear energy, utilization of heavy machinery may not be possible.

Each mine in practice requires settlements of people to work the mines. Figure 14.4 shows a possible lunar settlement. Although this type design may seem to be attractive, the feasibility of such settlements for the people, who would be working for, may be problematic.

FIGURE 14.3 An illustration of an open-pit mine for lunar mining activities (from Shutterstock).

FIGURE 14.4 A possible lunar settlement (from NASA).

14.3 CONSTRUCTION

Any mining activity would definitely require the establishment of settlements for communities. This requirement would entail the construction of buildings, roads, bridges and tunnels (Figures 14.5 and 14.6). Construction would probably involve concrete mixing and casting, reinforcement with the consideration of gravity, as well as seismic loads. Quake observation on the Moon and Mars has shown that the shaking duration is quite large and may have unknown effects on structures. Another issue might be waste management resulting from constructions as well as human activities. Furthermore, making concrete under airless conditions or in atmosphere consisting of only CO_2 may create other problems and necessitate the utilization of special cements, which can harden under such circumstances.

FIGURE 14.5 Artist illustration of a construction on the Moon (from NASA).

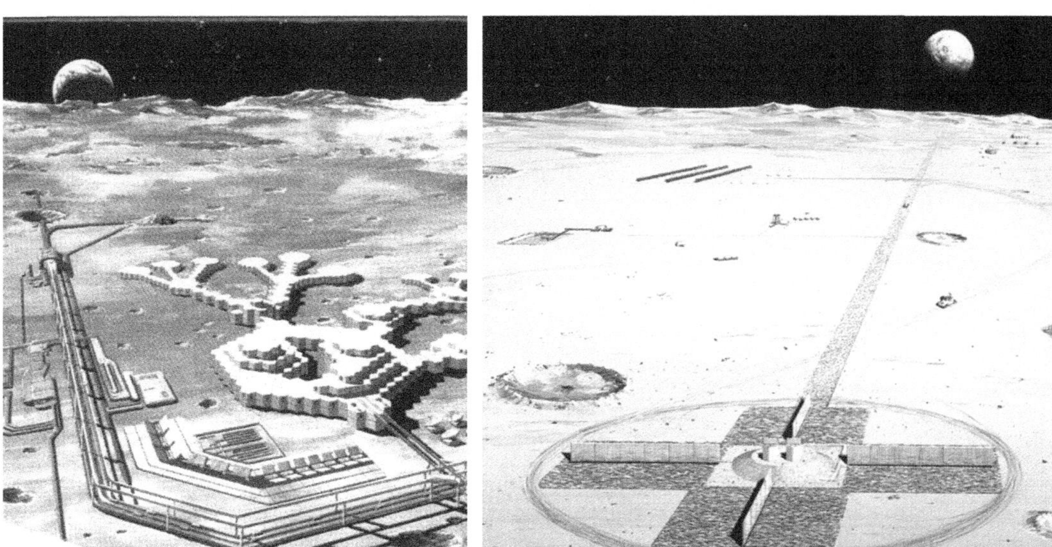

FIGURE 14.6 Possible construction of lunar bases and power plant (courtesy of Shimizu Co.).

14.4 EXCAVATION

As mentioned in previous subsections, mining and settlement construction would definitely involve excavations. The dust created during excavation in the Moon and Mars might be quite problematic. Such a problem was reported during the landing of the Apollo 15 lunar landing module, as the astronauts could not see the ground due to dust caused by the engine jets. Except for this issue, the stability issues during excavations should be similar to those on Earth.

14.5 ENVIRONMENTAL ISSUES

The Moon lacks atmosphere and provides no protection from space radiation or meteoroids. Therefore, the utilization of housing partially covered by soil or lunar lava tubes has been proposed to act as a shield against radiation, as illustrated in Figure 14.7.

FIGURE 14.7 Artist's illustration of a lunar housing (from ESA).

As discussed in Chapter 10, the skylights (sinkholes, pits) are considered potential sites of settlement for humans. Figure 14.8 show an underground café in Gangala natural cave of Okinawa Island as an example of the use of an underground cavity on Earth. As the gravity of the Moon and Mars are lower than that of Earth, the span of stable underground openings may be much greater than those on Earth. The span of the largest karstic cave on Earth is 300 m. However, it should be noted that the discontinuities in rock mass play negative roles for roof stability of underground openings. Some simple elastic analyses of caves on the Moon suggested that the span can be up to 5000 m. Nevertheless, conclusions from such a simple analysis should not be taken for granted. As the width of the largest collapsed pit named Copernicus LB3 on the Moon is only 430 m, it would not be appropriate to say that the Moon lava caves cannot support spans greater than 430 m with the consideration of rock mass conditions as they would have discontinuities resulting from cooling, successive eruptions and tectonic movements.

FIGURE 14.8 Utilization of a natural cave as a café on Okinawa Island of Earth.

The Moon's low surface gravity is also another concern. It is unknown whether 1/6 g is enough to maintain human health and human bones for long periods.

References

Aitta, A. (2012). Venus' internal structure, temperature and core composition. *Icarus*, 218, 967–974.

Aki, K. (1966). Generation and propagation of G waves from the Niigata earthquake of June 14, 1964. Part 2. Estimation of earthquake moment, released energy and stress-strain drop from G wave spectrum. *Bulletin of the Earthquake Research Institute*, 44, 73–88.

Alejano, L., Muralha, J., Ulusay, R., Li, C., Pérez-Rey, I., Karakul, H., Chryssanthakis, P., & Aydan, Ö. (2018). ISRM suggested method for determining the basic friction angle of planar rock surfaces by means of tilt tests. *Rock Mechanics and Rock Engineering*, 51(12), 3853–3859.

Amadei, B., & Stephanson, O. (1997). *Rock stress and its measurement*. Chapman & Hall.

Ambraseys, N. N. (1988). Engineering seismology. *Journal of Earthquake Engineering & Structural Dynamics*, 17, 1–105.

Amorim, D. O., & Gudkova, T. V. (2023). Internal structure of Venus based on the PREM model. *Solar System Research*, 57(5), 414–425.

Anderson, D. L., Miller, W. F., Latham, G. V., Nakamura, Y., Toksöz, M. N., Dainty, A. M., Duennebier, F. K., Lazarewicz, A. R., Kovach, R. L., & Knight, T. C. D. (1977). Seismology on Mars. *Journal of Geophysical Research*, 82(28), 4524–4546.

Anderson, L. D., & Hart, R. S. (1976). An earth model based on free oscillations and body waves. *Journal of Geophysical Research*, 81, 1461–1475.

Ando, M., Tsunoda, F., Hayakawa, Y., Hirahara, K., & Fujita, Y. (1996). *Earthquake and Volcanoes* (Jishin to Kazan—in Japanese, p. 191). Tokai University Publishing Co.

Attree, N., Groussin, O., Jorda, L., et al. (2018). Tensile strength of 67P/Churyumov–Gerasimenko nucleus material from overhangs. *A&A*, 611, A33.

Aydan, Ö. (1989). *The stabilisation of rock engineering structures by rockbolts*. Doctorate Thesis, Nagoya University, Faculty of Engineering.

Aydan, Ö. (1993). A consideration on the stress state of the earth due to the gravitational pull. *The 7th Annual Symposium on Computational Mechanics*, Tokyo, pp. 243–252.

Aydan, Ö. (1995). The stress state of the earth and the earth's crust due to the gravitational pull. *The 35th US Rock Mechanics Symposium*, Lake Tahoe, pp. 237–243.

Aydan, Ö. (1996). Faulting and characteristics of earthquake waves in Hyogo-ken Nanbu Earthquake of January 17, 1995 (in Turkish). *Jeoloji Mühendisliği*, 48, 63–77.

Aydan, Ö. (1997). *The seismic characteristics and the occurrence pattern of Turkish earthquakes*. Turkish Earthquake Foundation, Report No.: TDV/TR 97-007, 41p.

Aydan, Ö. (2000a). A stress inference method based on structural geological features for the full-stress components in the earth's crust. *Yerbilimleri*, 22, 223–236.

Aydan, Ö. (2000b). A stress inference method based on GPS measurements for the directions and rate of stresses in the earth' crust and their variation with time. *Yerbilimleri*, 22, 21–32.

Aydan, Ö. (2001). Comparison of suitability of submerged tunnel and shield tunnel for subsea passage of Bosphorus. *Geological Engineering Journal*, 25(1), 1–17.

Aydan, Ö. (2002). The inference of the earthquake fault and strong motions for Kutch earthquake of January 26, 2001. A symposium on the records and lessons from the recent large domestic and overseas earthquakes. *JSCE, Earthquake Engineering Committee*, Tokyo, pp. 135–140.

Aydan, Ö. (2003). Actual observations and numerical simulations of surface fault ruptures and their effects engineering structures. *The Eight U.S.-Japan Workshop on Earthquake Resistant Design of Lifeline Facilities and Countermeasures against Liquefaction*. Technical Report, MCEER-03-0003, pp. 227–237.

Aydan, Ö. (2006). Geological and seismological aspects of Kashmir Earthquake of October 8, 2005 and geotechnical evaluation of induced failures of natural and cut slopes. *Journal of Marine Science and Technology, Tokai University*, 4(1), 25–44.

Aydan, Ö. (2007). Inference of seismic characteristics of possible earthquakes and liquefaction and landslide risks from active faults (in Turkish). *The 6th National Conference on Earthquake Engineering of Turkey*, Istanbul, Vol. 1, pp. 563–574.

Aydan, Ö. (2008). Investigation of the seismic damage to the cave of Gunung Sitoli (Tögi-Ndrawa) by the 2005 Great Nias earthquake. *Yerbilimleri*, 29(1), 1–16.

Aydan, Ö. (2009). The 13 March 1992 Erzincan Earthquake and Some Thoughts and Evaluations on its Surface Effects. *Bulletin of Engineering Geology, Turkey*, 28–29, 33–56.

Aydan, Ö. (2012a). Ground motions and deformations associated with earthquake faulting and their effects on the safety of engineering structures. In R. Meyers (Ed.), *Encyclopedia of Sustainability Science and Technology* (pp. 3233–3253). Springer.

Aydan, Ö. (2012b). The inference of physico-mechanical properties of soft rocks and the evaluation of the effect of water content and weathering on their mechanical properties from needle penetration tests. *ARMA 12-639, 46th US Rock Mechanics/Geomechanics Symposium*, Chicago, Paper No. 639, 10 pages (on CD).

Aydan, Ö. (2015). Large rock slope failures induced by recent earthquakes. *Rock Mechanics and Rock Engineering, Special Issue on Deep-seated Landslides*, 49(6), 2503–2524.

Aydan, Ö. (2016). An integrated approach for the evaluation of measurements and inferences of in-situ stresses. *RS2016 Symposium, 7th International Sym*posium *on In-situ Rock Stress*, Tampere, pp. 38–57.

Aydan, Ö. (2017a). *Time dependency in rock mechanics and rock engineering*. CRC Press, Taylor and Francis Group, 241p.

Aydan, Ö. (2017b). *Rock dynamics*. CRC Press, Taylor and Francis Group, 462p. ISRM Book Series No. 3, ISBN 9781138032286.

Aydan, Ö. (2017c). *Some thoughts about rock mechanics aspects of Mars*. UNSW, 3rd Off Earth Mining Forum—2017OEMF.

Aydan, Ö. (2018a). Multiparameter and infrared monitoring systems for real-time rockburst susceptibility evaluation and their applications to practice in Japan. In Xia-Ting Feng (Ed.), Sub-Chapter 8.4, *Rockburst: Mechanisms, monitoring, warning and mitigation* (pp. 282–299). Elsevier.

Aydan, Ö. (2018b). Some thoughts on the risk of natural disasters in Ryukyu archipelago. *International Journal of Environmental Science and Development*, 9(10), 282–289.

Aydan, Ö. (2019). Dynamic shear tests on rock discontinuities and some considerations. *Proceedings of ISRM 14th International Congress of Rock Mechanics—Foz do Iguassu*, Brazil, pp. 957–964.

Aydan, Ö. (2020a). *Rock mechanics and rock engineering: Volume 1—Fundamentals of rock mechanics* (p. 31). CRC Press, Taylor and Francis Group, 379p. ISBN 978-0-367-42162-5.

Aydan, Ö. (2020b). *Rock mechanics and rock engineering: Volume 2—Applications*. CRC Press, Taylor and Francis Group, 410p.

Aydan, Ö. (2020c). Some thoughts on rocks and Rock Engineering in Mars (in Turkish). *Yer Mühendisliği*, 8(14), 38–46.

Aydan, Ö. (2021a). Some thoughts on Rock Mechanics in Mars. *15th Japan Domestic Rock Mechanics Symposium*, Paper No. 49, 273–278.

Aydan, Ö. (2021b). A Thermo-elasto-plastic Evaluation of Stress State of Mars. *EUROCK2021, IOP Conference Series: Earth and Environmental Science*, 833, 012089, 8p.

Aydan, Ö. (2021c). *Continuum and computational mechanics for geomechanical engineers*. CRC Press, Taylor and Francis Group, 324p. ISRM Book Series No. 7, ISBN 978036768053.

Aydan, Ö. (2022a). Some thoughts on rock slope stability issues in Mars. *EUROCK2022, Helsinki*, 8p.

Aydan, Ö. (2022b). Some considerations on rock dynamic issues in Mars. *AusRock2022, Melbourne*, 43–51.

Aydan, Ö. (2022c). A comparative study on the determination of tensile strength of rocks by different testing methods and its utilization in failure criteria. *ROCKMEC'2022–13th Regional Rock Mechanics Symposium*, Isparta, Turkey, pp. 3–11.

Aydan, Ö. (2022d). *Sinkhole Issues in Mars*. IX Latin America Rock Mechanics Symposium, Asuncion, Paraguay, 10p.

Aydan, Ö. (2023a). *Earthquake science and engineering*. CRC Press, Taylor and Francis Group, 496p. ISBN 9780367758776.

Aydan, Ö. (2023b). Some rock dynamics issues in Mars. *49th Rock Mechanics Symposium of Japan*, pp. 36–41.

Aydan, Ö., Akagi, T., & Kawamoto, T. (1996b). The squeezing potential of rock around tunnels: Theory and prediction with examples taken from Japan. *Rock Mechanics and Rock Engineering*, 29(3), 125–143.

Aydan, Ö., Akagi, T., Okuda, H., & Kawamoto, T. (1994). The cyclic shear behaviour of interfaces of rock anchors and its effect on the long term behaviour of rock anchors. *International Symposium on New Developments in Rock Mechanics and Rock Engineering*, Shenyang, 15–22.

Aydan, Ö., & Hamada, M. (1992). The site investigation of the Erzincan (Turkey) Earthquake of March 13, 1992. *4th Japan-US Workshop on earthquake Resistant Design of Lifeline Facilities and Countermeasures against Soil Liquefaction*, Honolulu, May, pp. 17–34.

Aydan, Ö., Hamada, M., & Sakamoto, A. (2007a). Characteristics of Noto Peninsula earthquakes with an emphasis on induced geotechnical damage. *Journal of the School of Marine Science and Technology*, 5(2), 13–31.

Aydan, Ö., Hamada, M., & Suzuki, Y. (2005a). Some observations and considerations on the damage induced by the tsunami of the 2004 Sumatra earthquake on structures and coast. *Journal of the School of Marine Science and Technology, Tokai University*, 3(1), 79–94.

Aydan, Ö., Ichikawa, Y., Shimizu, Y., & Murata, K. (1991). An integrated system for the stability of rock slopes. *The 5th International Conference on Computer Methods and Advances in Geomechanics, Cairns*, 1, 469–465.

Aydan, Ö., Kumsar, H., & Ulusay, R. (2002). How to infer the possible mechanism and characteristics of earthquakes from the striations and ground surface traces of existing faults. *JSCE, Earthquake and Structural Engineering Division*, 19(2), 199–208.

Aydan, Ö., & Kawamoto, T. (1992). The stability of slopes and underground openings against flexural toppling and their stabilisation. *Rock Mechanics and Rock Engineering*, 25(3), 143–165.

Aydan, Ö., & Kawamoto, T. (1994). The stress state of the interior and the crust of the earth (in Japanese). *The 9th National Symposium in Rock Mechanics of Japan,* Tokyo, pp. 635–640.

Aydan, Ö., & Kawamoto, T. (2000). Assessing mechanical properties of rock masses by RMR rock classification method. *GeoEngineering 2000 Symposium, Sydney*, OA0926.

Aydan, Ö., & Kim, Y. (2002). The inference of crustal stresses and possible earthquake faulting mechanism in Shizuoka Prefecture from the striations of faults. *Journal of the School of Marine Science and Technology, Tokai University*, 54, 21–35.

Aydan, Ö., Kiyota, R., Iwata, H., & Malistani, N. (2023). Tilting and stick–slip tests for evaluating static and dynamic frictional properties of rock discontinuities. *Rock Mechanics and Rock Engineering*, 56, 8607–8622.

Aydan, Ö., & Kumsar, H. (1997). *A site investigation of Oct. 1, 1995 Dinar Earthquake.* Turkish Earthquake Foundation, TDV/DR 97-003, 115p.

Aydan, Ö., & Kumsar, H. (2005, September 21–25). An Evaluation of Kuşini Antique Marble Quarry in İzmir, Türkiye as an underground rock engineering structure, *Proceedings of the 1st International Symposium on Travertine*, Denizli, Turkey, pp. 269–276.

Aydan, Ö., & Kumsar, H. (2010). An experimental and theoretical approach on the modeling of sliding response of rock wedges under dynamic loading. *Rock Mechanics and Rock Engineering*, 43(6), 821–830.

Aydan, Ö., Mamaghani, I. H. P., & Kawamoto, T. (1996a). Application of discrete finite element method (DFEM) to rock engineering structures. *NARMS'96*, 2039–2046.

Aydan, Ö., Miwa, S., Kodama, H., & Suzuki, T. (2005b). The characteristics of M8.7 Nias earthquake of March 28, 2005 and induced tsunami and structural damages. *Journal of the School of Marine Science and Technology, Tokai University*, 3(2), 66–83.

Aydan, Ö., Miwa, S., Kodama, K., Suzuki, T., & Hamada, M. (2007b). Support activities of JSCE and EWoB-Japan for Nias Island following the Great Nias earthquake of 2005. *The International Symposium on Disaster in Indonesia (ISDI): Problems and Solutions*, MS-3, Padang, pp. 1–20.

Aydan, Ö., & Nawrocki, P. (1998). Rate-dependent deformability and strength characteristics of rocks. *International Symposium on the Geotechnics of Hard Soils-Soft Rocks, Napoli*, 1, 403–411.

Aydan, Ö., & Paşamehmetoğlu, A. G. (1994). Dünyanın çeşitli yörelerinde ölçülmüş yerinde gerilimler ve yatay gerilim katsayısı (In-situ stress measurements in various parts of Earth and lateral stress coefficient, in Turkish). *Kaya Mekaniği Bülteni*, 10, 1–17.

Aydan, Ö., Sato A., & Yagi. M. (2014a). The inference of geo-mechanical properties of soft rocks and their degradation from needle penetration tests. *Rock Mechanics and Rock Engineering*, 47, 1867–1890.

Aydan, Ö., Sezaki, M., & Yarar, R. (1996e). The seismic characteristics of Turkish Earthquakes. *11th World Conference on Earthquake Engineering*, CD-2, Paper No 1025.

Aydan, Ö., Shimizu, Y., & Ichikawa, Y. (1989). The effective failure modes and stability of slopes in Rock Mass with Two Discontinuity Sets. *Rock Mechanics and Rock Engineering*, 22(3), 163–188.

Aydan, Ö., Shimizu, Y., & Kawamoto, T. (1992a). The stability of rock slopes against combined shearing and sliding failures and their stabilisation. *International Symposium on Rock Slopes, New Delhi*, 203–210.

Aydan, Ö., Shimizu, Y., & Kawamoto, T. (1992b). The reach of slope failures. *The 6th International Symposium on Slope failures, ISL 92*, Christchurch, 1, 301–306.

Aydan, Ö., & Tokashiki, N. (2007). Some damage observations in Ryukyu limestone caves of Ishigaki and Miyako Islands and their possible relations to the1771 Meiwa Earthquake. *Journal of the School of Marine Science and Technology, Tokai University*, 5(1), 23–39.

Aydan, Ö., & Tokashiki, N. (2011). A comparative study on the applicability of analytical stability assessment methods with numerical methods for shallow natural underground openings. *The 13th International Conference of the International Association for Computer Methods and Advances in Geomechanics,* Melbourne, Australia, pp. 964–969.

Aydan, Ö., Tokashiki, N., & Geniş, M. (2012). Some considerations on yield (failure) criteria in rock mechanics ARMA 12-640. *Proceeding of 46th US Rock Mechanics/Geomechanics Symposium*, Chicago, 10p, (on CD).

Aydan, Ö., Tokashiki, N., Kumsar, H., & Geniş, M. (2017, March 20). Some empirical guidelines on the stability of karstic caves based on Rock Mass Quality Rating (RMQR) classification system. *1st International Workshop on Role of Microorganism and Thermo-Hydro-Chemico-Mechanical Process in Geoengineering and Geoscience*, Okinawa, Japan, pp. 43–50.

Aydan, Ö., Üçpırtı, H., & Kumsar, H. (1996c). Stability of slopes having an inclined sliding surface with a visco-plastic behaviour. *Korea-Japan Joint Symposium on Rock Engineering*, 437–444.

Aydan, Ö., Üçpırtı, H., & Kumsar, H. (1996d). The stability of a rock slope having a visco-plastic sliding surface. *Kaya Mekaniği Bülteni*, 12, 39–49.

Aydan, Ö., & Ulusay, R. (2002). Back analysis of a seismically induced highway embankment during the 1999 Düzce earthquake. *Environmental Geology*, 42, 621–631.

Aydan, Ö., & Ulusay, R. (2003). Geotechnical and geoenvironmental characteristics of man-made underground structures in Cappadocia, Turkey. *Engineering Geology*, 69, 245–272.

Aydan, Ö., & Ulusay, R. (2013). Geomechanical evaluation of Derinkuyu Antique Underground City and its implications in geoengineering. *Rock Mechanics and Rock Engineering*, 46, 731–754.

Aydan, Ö., & Ulusay, R. (2024, 6–8 June). Interrelations among various indices and physico-mechanical properties of tuffs of Cappadocia Region of Türkiye. *ENGGEO'2024: National Symposium on Engineering Geology and Geotechnics*, NEVÜ, Nevşehir, pp. 73–83.

Aydan, Ö., Ulusay, R., Hasgür, Z., & Taşkın, B. (2000a). A site investigation of Kocaeli Earthquake of August 17, 1999. *Turkish Earthquake Foundation*, 08–49, 180.

Aydan, Ö., Ulusay, R., Kumsar, H., & Ersen, A. (1996f). Buckling failure at an open-pit coal mine. *EUROCK'96*, 641–648.

Aydan, Ö., Ulusay, R., Kumsar, H., & Tuncay, E. (2000b). *Site investigation and engineering evaluation of the Düzce-Bolu Earthquake of November 12, 1999.* Turkish Earthquake Foundation, TDV/DR 09-51, 220p.

Aydan, Ö., Ulusay, R., & Tokashiki, N. (2014b). A new rock mass quality rating system: Rock Mass Quality Rating (RMQR) and its application to the estimation of geomechanical characteristics of rock masses. *Rock Mech Rock Engineering*, 47, 1255–1276.

Aydan, Ö., Watanabe, S., & Tokashiki, N. (2008). The inference of mechanical properties of rocks from penetration tests. *Proceedings of the 5th Asian Rock Mechanics Symposium (ARMS5)*, Tehran, pp. 213–220.

Aydan, T., Akagi, T., & Kawamoto, T. (1993). Squeezing potential of rocks around tunnels; theory and prediction. *Rock Mechanics and Rock Engi*neering, 26(2), 137–163.

Barton, N. (1976). Unsupported under-ground openings. In *Rock Mechanics Meeting, Swedish Rock Mechanics research Foundation* (pp. 61–94). Stockholm.

Barton, N. R., Lien, R., & Lunde, J. (1974). Engineering classification of rock masses for the design of tunnel support. *Rock Mechanics*, 6(4), 189–239.

Barton, N., Løset, F., Lien, R., & Lunde, J. (1980). Application of the Q-system in design decisions. In M. Bergman (Ed.), *Subsurface space* (Vol. 2, pp. 553–561). New York: Pergamon.

Baslilevsky, A. T., et al. (2017). Geologic analysis of the Rosetta NavCam, Osiris and ROLIS images of the comet 67P/Churyumov-Gerasimenko nucleus. *Planetary and Space Science*, 137, 1–19.

Bickel, V. T., Aaron, J., Manconi, A., & Loew, S. (2021). Global drivers and transport mechanisms of lunar rockfalls. *Journal of Geophysical Research: Planets*, 126(10), e2021JE006824.

Bieniawski, Z. T. (1973). Engineering classification of jointed rock masses. *Transactions of the South African Institution of Civil Engineers*, 15, 335–344.

Bieniawski, Z. T. (1989). *Engineering rock mass classifications.* New York: Wiley.

Brinkman, N., Stähler, S., Giardini, D., Schmelzbach, C., Khan, A., Jacob, A., Fuji, N., Perrin, C., Lognonné, P., Beucler, E., Böse, M., Ceylan, S., Charalambous, C., Clinton, J., Euchner, F., Horleston, A., Kawamura, T., Knapmeyer-Endrun, B., & Banerdt, W. (2021). First focal mechanisms of marsquakes. *Journal of Geophysical Research: Planets*, 126. https://doi.org/10.1029/2020JE006546.

Brissaud, Q., Krishnamoorthy, S., Jackson, J. M., Bowman, D. C., Komjathy, A., Cutts, J. A., Zhan, Z., Pauken, M. T., Izraelevitz, J. S., Gerald, J., & Walsh, G. J. (2021). The first detection of an earthquake from a balloon using its acoustic signature. *Geophysical Research Letters*. https://doi.org/10.1029/2021GL093013.

Brunetti, M. (2014). *Statistics of terrestrial and extraterrestrial landslides*. Doctoral Thesis, Universita Degli Studi di Perugia, 109p. https://doi.org/10.13140/2.1.4107.3444

Ceylan, S., Giardini, D., Clinton, J. F., Kim, D., Khan, A., Stähler, S. C., et al. (2023). Mapping the seismicity of Mars with InSight. *Journal of Geophysical Research: Planets*, 128, e2023JE007826.

Christiansen, E. H., & Spilker, B. (2018). *Exploring the planets*. Published by Prentice Hall. http://explanet.info

Çığ, M. İ. (2020). *Sümerler Türklerin bir koludur* (Sumers are a branch of Turks). Published by Kaynak Yayınları, 280p.

Collings, G. S., Melosh, J. J., & Marcus, R. A. (2005). Earth impact effects program: A web-based computer program for calculating the regional environmental consequences of a meteoroid impact on Earth. *Meteoritics & Planetary Science*, 40(6), 817–840.

Conway, S. J., Balme, M. R., Lamb, M. P., Towner, M. C., & Murray, J. B. (2011). Enhanced runout and erosion by overland flow under subfreezing and low pressure conditions: Experiments and application to Mars. *Icarus*, 211(1), 443–457.

Crosta, G. B., Utili, S., Blasio, F. V., & Riccardo Castellanza, R. (2014). Reassessing rock mass properties and slope instability triggering conditions in Valles Marineris, Mars. *Earth and Planetary Science Letters*, 388, 329–342.

Cruikshank, D. P., & Wood, C. A. (1971). Lunar rilles and Hawaiian volcanic features: Possible analogues. *The Moon*, 3, 412–447.

Cushing, G. E. (2012). Candidate cave entrances on Mars. *Journal of Cave and Karst Studies*, 33–47.

Cushing, G. E., Okubo, C. H., & Titus, T. N. (2015). Atypical pit craters on Mars: New insights from THEMIS, CTX, and HiRISE observations. *Journal of Geophysical Research: Planets*, 120, 1023–1043.

Cushing, G. E., Titus, T. N., Wynne, J. J., & Christensen, P. R. (2007). THEMIS observes possible cave skylights on Mars. *Geophysical Research Letters*, 34, L17201, 5p.

Dainthy, A. M., Toksöz, M. N., & Stein, S. (1976). Seismic investigation of the lunar interior. *Proceedings of the 7th Lunar Science Conference*, pp. 3057–3075.

Dobry, R., Idriss, I. M., & Ng, E. (1978). Duration characteristics of horizontal components of strong motion earthquake records. *Bulletin of the Seismological Society of America*, 68(5), 1487–1520.

Drucker, D. C., & Prager, W. (1952). Soil mechanics and plastic analysis for limit design. *Quarterly of Applied Mathematics*, 10 (2), 157–165.

Emre, Ö., Duman, T. Y., Doğan, A., Ateş, S., Keçer, Mç, Erkal, T., Özalp, S., Yıldırım, N., & Güner, N. (1999). *12 Kasım 1999 Düzce Depremi saha gözlemleri ve ön değerlendirme raporu*. MTA.

Erentöz, C. (1974). *Türkiye Jeloji Haritası (Denizli), 1/500000 ölçekli*. MTA Yayınları, Ankara.

Ergin, K., Güçlü, U. ve Uz, Z., (1967). *The catalogue of earthquakes of Turkey—Türkiye ve civarının deprem kataloğu (MS. 11-1964)*. İTÜ Maden Fakültesi Arz Fiziği Enstitüsü Yayınları, No 28 (in Turkish and English).

Erguvanlı, K. (1973). *Engineering geology* (in Turkish). Istanbul: ITU Press. No. 966., 552p.

Eringen, A. C. (1980). *Mechanics of continua*. Huntington, NY: Robert E. Krieger Publishing Co., 1980, 606p.

Europe Space Agency (ESA). (2019). *Mars Express*. https://sci.esa.int/

Eyidoğan, H., Güçlü, U., Utku, Z., & Değirmenci, E. (1991). *A macro-seismic catalog of great earthquakes of Turkey for 1900–1988* (Türkiye büyük depremleri makrosismik rehberi), İTÜ, 199p.

Fowler, C. M. R. (1990). *The solid earth—An introduction to Global Geophysics*. Cambridge: Cambridge University Press.

Gençoğlu, S., İnan, I., & Güler, H. (1990). *Türkiyenin Deprem Tehlikesi* (Earthquake danger of Turkey). Ankara: Publication of the Chamber of Geophysical Engineers of Turkey, 701p.

Groussin, O., Jorda, L., Auger, A.-T., et al. (2018). Gravitational slopes, geomorphology, and material strengths of the nucleus of comet 67P/Churyumov-Gerasimenko from OSIRIS observations. *Astronomy & Astrophysics*, manuscript no. 26379_ap.

Gudkova, T. V., & Zharkov, V. N. (2020). Models of the internal structure of the Earth-like Venus. *Solar System Research*, 54(1), 20–27.

Gürpınar, A., Erdik, M., Yücemen, S., & Öner, M. (1979). Risk Analysis of Northern Anatolia Based on Intensity Attenuation. *Proceedings of the 2nd U.S. National Conference on Earthquake Engineering*, Stanford, California, pp. 72–81.

Hanks, T. C., & Kanamori, (1979). A moment magnitude scale. *Journal of Geophysical Research*, 84(B5), 2348–50.

Haruyama, J., Morota, T., Kobayashi, S., Sawai, S., Lucey, P. G., Shirao, M., & Nishino, M. N. (2012). Lunar holes and lava tubes as resources for lunar science and exploration. In V. Badescu (Ed.), *Moon: Prospective energy and material resources* (pp. 139–163). Heidelberg, Germany: Springer.

Heather, H. (1999). *Ulughbek and his observatory in Samarkand*. Englan: Cosmic Elk.

Herzfeld, E. (1930). *Die vorgeschichtlichen Töpfereien von Samarra. Die Ausgrabungen von Samarra* (Vol. 5). Berlin: Dietrich Reimer/Ernst Vohsen.

Hirth, G., & Tullis, J. (1994). The brittle-plastic transition in experimentally deformed quartz aggregates. *Journal of Geophysical Research*, 99, 11731–11747.

Hoek, E., & Brown, E. T. (1980). Empirical strength criterion for rock masses. *Journal of the Geotechnical Engineering Division*, ASCE, 106(GT9), 1013–1035.

Hoek, E., & Brown, E. T. (1997). Practical estimates of rock mass strength. *International Journal of Rock Mechanics and Mining Sciences*, 34(8), 1165–1186.

Horiuchi, K., Aydan, Ö., Tokashiki, N., & Nasiry, N. Z. (2019). Some considerations on the causes of cliff failures in Ryukyu limestone formation in Ryukyu Archipelago and the analysis of a recent cliff failure. *JSCE, Division C: Geotechnics*, 75(2), 216–233.

IRIS. (2013). *Teachable moments. M0.0 meteor explosion, Russia: Friday, February 15, 2013 at 03:20:26 UTC.* https//www.iris.edu/130215Russia.

Irvine, T. (2002, March). Moonquakes. *Newsletter, Vibrationdata.com*, 4–13.

ISRM. (2007). The complete ISRM suggested methods for rock characterization, testing and monitoring: 1974–2006 (R. Ulusay, Ed.), ISRM Turkish National Group, 628p.

Jeffreys, H., & Bullen, K. E. (1940). *Seismological Tables*. London: British Association for the Advancement of Science.

Johnston, D. H., & Toksöz, M. N. (1977). Internal structure and properties of Mars. *Icarus*, 32, 73–84.

Kanamori, H. (1983). Global seismicity. In H. Kanamori & E Boschi (Eds.), *Earthquakes: Observation, theory and interpretation* (pp. 596–608). New York: North-Holland.

Kawamoto, T., Aydan, Ö., & Tsuchiyama, S. (1991). A consideration on the local instability of large underground openings. *Proceedings of the International Conference GEOMECHANICS'91*, Hradec, pp. 33–41.

Kawamoto, T., & Aydan, Ö. (1999). A review of numerical analysis of tunnels in discontinuous rock masses. *International Journal of Numerical and Analytical Methods in Geomechanics*, 23, 1377–1391.

Kenkmann, T., Wunnemann, K., Deutsch, A., Poelchau, M. H., Schafer, F., & Thoma, K. (2011). Impact cratering in sandstone: The MEMIN pilot study on the effect of pore water. *Meteoritics & Planetary Science*, 46, 890–902.

Ketin, İ. (1973). *Umumi Jeoloji* (General geology). Istanbul: Published by İ. T. Ü. No: 30.

Kreslavsky, M. A., Zharkova, A. Y., Head, J. W, Maria, I., & Gritsevich, M. I. (2021). Mercury boulders. *Icarus*, 369, 114628.

Kudo, K. (1983). Seismic source characteristics of recent major Earthquakes in Turkey. In Y. Ohta (Ed.), A comprehensive study on Earthquake disasters in Turkey in view of seismic risk reduction (pp. 23–66). Sapporo, Japan: Hokkaido University.

Kumsar, H., & Aydan, Ö. (2022). An integrated study on the stability assessment and partial collapse of the Kaklık karstic cave (Denizli, Turkey). *Bulletin of Engineering Geology and the Environment*, 80(10), 221–230.

Kumsar, H., Aydan, Ö., Matsubara, H., Sakamoto, I., & Mori, K. (2023). Characteristics of hydro-magnesite of Salda Lake (Türkiye) and their implications on rocks of Jezerro Crater in Mars. *15th ISRM Congress*, Salzburg, pp. 882–887.

Kumsar, H., Aydan, Ö., & Ulusay, R. (2000). Dynamic and static stability of rock slopes against wedge failures. *Rock Mechanics and Rock Engineering*, 33(1), 31–51.

Lang, B. (1994). *Span design for entry type excavations*. M.Sc. Thesis, British Columbia University, Vancouver, BC.

Latham, G., Ewing, M., Press, F., Sutton, G., Dorman, J., Nakamura, Y., Lammlein, D., Duennebier, F., & Toksöz, N. (1971). Moonquakes and lunar tectonism. *Science*, 174, 687–692.

Lindsay, H. (2008). *ALSEP Apollo lunar surface experiments package 19 Nov 1969–30 Sept 1977*. https://www.hq.nasa.gov/alsj/HamishALSEP.html.

Lucchitta, B. K. (1979). Landslides in Vallis Marineris, Mars. *Journal of Geophysical Research*, 84, 8097–8113.

Lyttleton, R. A. (1969). On the internal structures of mercury and Venus. *Astrophysics and Space Science*, 5, 18–35.

Magnarini, G., Bell, S. K., Mitchell, T. M., Grindrod, P. M., Eckley, S., Joy, K. H., et al. (2023, June 27–29). Investigation of the Light Mantle landslide deposit emplacement dynamics through clast fabric and morphology in the Apollo 17 core sample 73002. *11th European Lunar Symposium*, Padua, Italy.

Malin, C. M. (1992). Mass movements on Venus: Preliminary results from Magellan Cycle I observations. *Journal of Geophysical Research*, 97(E10), 16337–16352.

Maruto Co. (1999). *Instructions for SH-70 penetrometer* (NANGAN-PENETOROKEI). Tokyo, 4p. (in Japanese).

Mase, G. (1970). *Theory and problems of continuum mechanics*. Schaum Outline Series. McGraw Hill Co., 230p.

Matsuda, T. (1981). Active faults and damaging earthquakes in Japan. Macroseismic zoning and precaution fault zones. In *Maurice Ewing Series* (Vol. 4, pp. 271–289). Washington: American Geophysical Union.

Melosh, H. J. (1989). *Impact cratering: A geologic process*. New York, NY: Oxford University Press, 245pp.

Melosh, H. J. (2011). *Planetary surface processes*. Cambridge: Cambridge University Press, 500 pp.

Melosh, H. J., & McKinnon, W. B. (1988). The tectonics of Mercury. In F. Vilas, C. R. Chapman & M. S. Matthews (Eds.), *Mercury* (pp. 374–400). Tucson: University of Arizona Press.

Michel, N. C., Hauck, S. A., Solomon, S. C., Phillips, R. J., Roberts, J. H., & Zuber, M. T. (2013). Dynamics and evolution of Mercury's interior as constrained by MESSENGER observations. *Journal of Geophysical Research*, 118(5), 1033–1044.

Miki, K. (1986). *Introduction to rock mechanics (in Japanese)*. Tokyo: Kajima Press, 317p.

Milne, J. (1898). *Seismology*. Kegan Paul, Trench, Trubner & Co., 320p.

Moore, H. J. (1971). Craters produced by missile impacts. *Journal of Geophysical Research*, 76(23), 5750–5755.

Nakamura, Y. (2003). New identification of deep moonquakes in the Apollo lunar seismic data. *Physics of the Earth and Planetary Interiors*, 139, 197–205.

Nakamura, Y., Dorman, J., Duennebier, F., Lammlein, D., & Latham, G. (1975, May–July). Lunar Science Institute, symposium on origin and evolution of the lunar regolith, Houston, Tex., Nov. 13–15, 1974. *The Moon*, 13, 57–66.

NASA. (1997). *Apollo mission: Astronomy picture of the day: Apollo 17—Boulder on the Moon*. https://apod.nasa.gov/apod/ap970905.html

NASA. (2008). *Exploration: NASA's plans to explore the moon, mars and beyond*. http://www.nasa.gov/

NASA. (2019). *Images*. https://mars.nasa.gov/mer/gallery/images.html

NASA. (2022). *Images*. https://mars.nasa.gov/mer/gallery/images.html

NASA/JPL-Caltech. (2022). *NASA's InSight records monster quake on Mars*. https://www.nasa.gov/missions/insight/nasas-insight-records-monster-quake-on-mars/

NASA-News. (2019). *NASA's InSight detects first likely 'quake' on Mars*. InSight Mission. https://mars.nasa.gov/news/8430/

Neuffer, P. D., & Schultz, R. A. (2006). Mechanisms of slope failure in Valles Marineris, Mars. *Quarterly Journal of Engineering Geology and Hydrogeology*, 39 (3), 227–240.

Nimmo, F., & Watters, T. R. (2004). Depth of faulting on Mercury: Implications for heat flux and crustal and effective elastic thickness. *Geophysical Research Letters*, 31, L02701.

Ohta, Y., & Aydan, Ö. (2004). An experimental study on ground motions and permanent deformation nearby faults. *Journal of the School of Marine Science and Technology*, 2(3), 1–12.

Ohta, Y., & Aydan, Ö. (2009). An experimental and theoretical study on stick-slip phenomenon with some considerations from scientific and engineering viewpoints of earthquakes. *Journal of the School of Marine Science and Technology*, 8(3), 53–67.

Ohta, Y., Matsuzawa, R., & Katayama, Y. (2025). A study on the damage characteristics of Kimachi sandstone induced by high-velocity impact of a hemispherical rigid projectile. *51th Japan Rock Mechanics Symposium*, Tokyo, pp. 78–83. (in Japanese with English abstract).

Omori, F. (1894). On the aftershocks of earthquake. *Journal of the College of Science, Imperial University of Tokyo*, 7, 111–200.

Osinki, G. R. (2006). The geological record of meteorite impacts. *40th ESLAB First International Conference on Impact Cratering in the Solar System*, Noordwijk, The Netherlands.

Owen, D. R. J., & Hinton, E. (1980). *Finite element in plasticity: Theory and practice*. Swansea: Pineridge Press Ltd.

Perzyna, P. (1966). Fundamental problems in viscoplasticity. *Advances in Applied Mechanics*, 9(2), 244–368.

Pinet, P. R. (1992). *Oceanography: An introduction to the planet*. Saint Paul: West Publishing Company.

Pirajno, F., & van Kranendonk, M. J. (2005). Review of hydrothermal processes and systems on Earth and implications for Martian analogues. *Australian Journal of Earth Sciences*, 52, 329–351.

Ramsay, J. G., & Huber, M. I. (1984). *The techniques of modern structural geology, Vol. 1: Strain analysis*. London: Academic Press, 306p.

Richter, C. F. (1935). An instrumental earthquake magnitude scale. *Bulletin of Seismological Society of America*, 25(1), 1–32.

Roelofse, F., & Saunders, I. (2013). A first report on meteor-generated seismic signals as detected by the SANSN. *South African Journal of Science*, 109(5/6), 6.

Ryabinin, Y. N., Beresnev, B. I., & Martinov, E. D. (1971). Mechanical properties and processes in solids under high pressure. *Journal of Geophysical Research*, 76(5).

Rzhiga, O. N. (1988). *New epoch in exploration of Venus (Radar Imaging from Venera 15 and 16)*. Moscow: Znanie.

Sagdeev, R. Z., & Moroz, V. I. (1982, July–August). Venera-13 and Venera-14. *Soviet Astronomy Letters*, 8, 209–211.

Sagdeev, R. Z., et al., (1986, March). The VEGA Venus balloon experiment. *Science*, 231(4744), 1407.

Sato, R. (1989). *Handbook on parameters of earthquake faults in Japan*. Tokyo: Kajima Publishing Co. (in Japanese).

Schultz, R. A. (2002). Stability of rock slopes in Valles Marineris, Mars. *Geophysical Research Letters*, 29, 1932.

Scott, D. H., & Carr, M. H. (1978). *Geologic map of Mars: U.S. Geological Survey Investigations*. Series I-1083, scale 1:25000000. http://pubs.er.usgs.gov/publication/i1083

Seiki, T., & Aydan, Ö. (2003). Deterioration of Oya Tuff and its mechanical property change as building stone. *Proceedings of International Symposium on Industrial Minerals and Building Stones*, Istanbul, Turkey, pp. 329–336.

Sezaki, M., Aydan, Ö., Ichikawa, Y., & Kawamoto, T. (1990). Mechanical properties of rock mass for the pre-design of tunnels by NATM using a rock mass data-base (in Japanese). *Journal of Civil Engineers of Japan, Construction Division*, 421-VI-13, 125–133.

Sezaki, M., Aydan, Ö., & Kawamoto, T. (1994). A consideration on ground response curve (in Japanese). *Journal of Japan Society of Civil Engineers, Geotechnical Division*, 499/III-28, 77–85.

Shimada, M. (1993). Two types of brittle fracture of silicate rocks and scale effect on rock strength: Their implications in the earth crust. *Proceedings of the Workshop on Scale Effects in Rock Masses*, Lisbon, pp. 55–62.

Sohl, F., & Spohn, T. (1997). The interior structure of Mars: Implications from SNC meteorites. *Journal of Geophysical Research: Planets*, 102 (E1), 1613–1635.

Solomon, S. C. (1976). Some aspects of core formation in Mercury. *Icarus*, 28, 509–521.

Somuncuoğlu, S. (2010). *Sibiryadan Anadoluya Taştaki Türkler* (Turks on Stones from Siberia to Anadolu). Istanbul: Published by AC Yapi.

Somuncuoğlu, S. (2011). Saymalıtaş gökyüzü atları (Saymalıtaş skyhorses). Istanbul: Published by AC Yapi.

Steinberger, B., Werner, S. C., & Torsvik, T. H. (2010). Deep versus shallow origin of gravity anomalies, topography and volcanism on Earth, Venus and Mars. *Icarus*, 207, 564–577.

Terzaghi, K. (1943). *Theoretical soil mechanics*. New York: Wiley.

Tokashiki, N. (2011). *Study on the engineering properties of Ryukyu limestone and the evaluation of the stability of its rock mass and masonary structures*. PhD thesis, Waseda University. (in Japanese with English abstract).

Tokashiki, N., & Aydan, Ö. (2010). The stability assessment of overhanging Ryukyu limestone cliffs with an emphasis on the evaluation of tensile strength of Rock Mass. *Journal of Geotechnical Engineering, JSCE*, 66(2), 397–406.

Toksöz, M. N. (1974). Geophysical data and interior of the Moon. *Annual Review of Earth and Planetary Sciences*, 2, 151–177.

Toksöz, M. N. (1975). Lunar and planetary seismology. *Reviews of Geophysics*, 13(3), 306–311.

Toksöz, M. N., Dainty, A. M., Solomon, S. C., & Anderson, K. R. (1974). Structure of the Moon. *Reviews of Geophysics*, 12(4), 539–567.

Toksöz, M. N., Goins, N. R., & Cheng, C. H. (1977). Moonquakes: Mechanisms and relation to tidal stresses. *Science*, 196, 979–981.

Toksöz, M. N., & Hsui, A. T. (1978). Thermal history and evolution of Mars. *Icarus*, 34(3), 537–547.

Toksöz, M. N., Press, F., Anderson, K., Dainty, A., Latham, G., Ewing, M., Dorman, J., Lammlein, D., Nakamura, Y., Sutton, G., & Duennebier, F. (1972). Velocity structure and properties of the lunar crust. *Moon*, 4, 490.

Tosi, N., Grott, M., Plesa, A.-C., & Breuer, D. (2013). Thermochemical evolution of Mercury's interior. *Journal of Geophysical Research: Planets*, 118, 2474–2487.

Udias, A. (1999). *Principles of seismology*. Cambridge: Cambridge University Press, 475p.

Ulusay, R., & Aydan, Ö. (2005). Characteristics and geo-engineering aspects of the 2003 Bingöl (Turkey) earthquake. *Journal of Geodynamics*, 40(2–3), 334–346.

Ulusay, R., Aydan, Ö., Erguler, Z. A., Ngan-Tillard, D. J. M., Seiki, T., Verwaal, W., Sasaki, Y., & Sato, A. (2014). ISRM suggested method for the needle penetration test. *Rock Mechanics and Rock Engineering*, 47, 1073–1085.

Ulusay, R., Aydan, Ö., Erken, E., Kumsar, H., Tuncay, E., & Kaya, Z. (2003a). *Site investigation and engineering evaluation of the Cay-Eber Earthquake of February 3, 2002*. Turkish Earthquake Foundation, TDV/DR 012-79, p. 213.

Ulusay, R., Aydan, Ö., Erken, E., Kumsar, H., Tuncay, E., & Kaya, Z. (2003b). Characteristics of the Cay-Eber earthquake and its evaluation in terms of local ground conditions. *Fifth National Conference on Earthquake Engineering, 26–30 May 2003*, Istanbul, Turkey, Paper No: AE-115.

Ulusay, R., Aydan, Ö., & Hamada, M. (2001). The behaviour of structures built on active fault zones: Examples from the recent earthquakes of Turkey. *A Workshop on Seismic-Fault Induced Failures—Possible Remedies for Damage to Urban Facilities*, Tokyo, Japan, pp. 1–26.

Ulusay, R., Aydan, Ö., & Hamada, M. (2002). The behavior of structures built on active fault zones: Examples from the recent earthquakes of Turkey. *Structural Engineering/Earthquake Engineering, JSCE*, 19(2), Special Issue, 149–167.

Ulusay, R., Ersoy, H., Sünnetçi, M. O., & Karahan, M. (2025). The leeb (Equotip) hardness test for rock materials: An overview, assessments on the factors influencing test results, and prediction models based on a large database. *Bulletin of Engineering Geology and the Environment*, 84, Article number 148.

Ulusay, R., Tuncay, E., Sonmez, H., & Gokceoglu, C. (2004). An attenuation relationship based on Turkish strong motion data and iso-acceleration map of Turkey. *Engineering Geology*, 74, 265–291.

USGS. (2020). *Geologic map of Mars*. https://pubs.usgs.gov/sim/3292/

Wagner, R. V., & Robinson, M. S. (2014). Distribution, formation mechanisms, and significance of lunar pits. *ICARUS*, 237, 52–60.

Watters, T. R. (1995). *Planets: A smithsonian guide*. USA: MacMillan, 256p.

Watters, T. R., James, P. B., & Selvans. (2002). M. M. Mercury's crustal thickness and contractional strain. *Geophysical Research Letters*, 48, e2021GL093528. https://doi.org/10.1029/2021GL093528

Waversik, W. R., & Fairhurst, C. (1970). A study of brittle rock fracture in laboratory compression experiments. *International Journal of Rock Mechanics and Mining Sciences & Geomechanics Abstracts*, 7, 561–575.

Weber, R. C., Lin, P. Y., Garnero, E. J., Williams, Q., & Lognonné, P. (2011). Seismic detection of the lunar core. *Science*, 331, 309–312.

Wells, D. L., & Coppersmith, K. J. (1994). New empirical relationship among magnitude, rupture length, rupture width, rupture area, and surface displacement. *Bulletin of the Seismological Society of America*, 84(4), 974–1002.

West, M. D., & Clarke, J. D. A. (2010). Potential Martian mineral resources: Mechanisms and terrestrial analogues. *Planetary and Space Science*, 58, 574–582.

Yarar, R., Ergunay, O., Erdik, M., & Gülkan, P. (1980). A preliminary probabilistic Assessment of the Seismic Hazard in Turkey. *Proceedings of the 7WCEE*, Istanbul, Vol. 1, pp. 309–316.

Yüzer, E., & Vardar, M. (1986). *Rock mechanics* (in Turkish). ITU Foundation, Istanbul. No. 4, 154p.

Zharkov, V., & Gudkova, T. (2005). Construction of Martian interior model. *Journal of Solar System Research*, 39(5), 343–373.

Index

For Product Safety Concerns and Information please contact our
EU representative GPSR@taylorandfrancis.com Taylor & Francis
Verlag GmbH, Kaufingerstraße 24, 80331 München, Germany